Urban
Planning
and the
African
American
Community

From June:

*To my parents, President Emeritus Hubert V. Manning,
and Associate Professor Emeritus Ethel A. Manning.*

From Marsha:

*To my husband, Paul V. Brozovsky,
and my son, Jon M. Ritzdorf.*

Urban Planning and the African American Community

In the Shadows

June Manning Thomas
Marsha Ritzdorf
Editors

SAGE Publications
International Educational and Professional Publisher
Thousand Oaks   London   New Delhi

BJL 3954 - 7/3

For information address:

SAGE Publications, Inc.
2455 Teller Road
Thousand Oaks, California 91320
E-mail: order@sagepub.com

SAGE Publications Ltd.
6 Bonhill Street
London EC2A 4PU
United Kingdom

SAGE Publications India Pvt. Ltd.
M-32 Market
Greater Kailash I
New Delhi 110 048 India

Printed in the United States of America

Library of Congress Cataloging-in-Publication Data

Main entry under title:
Urban planning and the African American community: In the shadows. Authors,
  June Manning Thomas and Marsha Ritzdorf.
      p.   cm.
  Includes bibliographical references and index.
  ISBN 0-8039-7233-4 (cloth: acid-free paper).—ISBN 0-8039-7234-2
(pbk.: acid-free paper).
    1. City planning—United States. 2. Afro-Americans—Social conditions.
  3. Urban renewal—United States. 4. Zoning—United States. 5. United States—
  Race relations. I. Thomas, June Manning. II. Ritzdorf, Marsha.
HT167.U7277 1997
307.1'216'08996073—dc20   96-25329

97 98 99 00 01 02 03 10 9 8 7 6 5 4 3 2 1

Acquiring Editor:	Catherine Rossbach
Editorial Assistant:	Nancy Hale
Production Editor:	Diana E. Axelsen
Production Assistant:	Karen Wiley
Typesetter/Designer:	Marion Warren
Cover Designer:	Ravi Balasuriya
Cover Photo:	Nemo Warr
Print Buyer:	Anna Chin

CONTENTS

ACKNOWLEDGMENTS

When we began this book, friends teased us, saying that once we had edited a volume we would never do it again! Well, they are wrong: we have an extraordinary set of responsive authors to thank for what has been an uplifting experience for both of us. We thank them all for their timely responses to our editing. We would work with them again anytime.

Integral to our survival have been the contributions of Fran Fowler of Michigan State University, Urban Affairs Programs, who provided the most critical help with the typing, revision, and overall production of this manuscript. We both thank her for her enormous services. Marsha would also like to thank Janet Town at Virginia Tech for secretarial support. Carrie Mullen, formerly of Sage, and Catherine Rossbach and Diana Axelsen, our current editors at Sage, deserve thanks for their help in shepherding the manuscript through the production process. Finally, we thank our family, friends, and associates, who supported us through this endeavor.

June would like to thank especially Marsha, a true sister at heart and soul if not in blood; Michigan State University, most especially the Urban

Affairs Programs, which provided extensive support; my graduate assistant, Vince Richardson; David L. Poremba, at the Burton Historical Collection, Detroit Public Library; and librarians at the Bohn and Municipal Reference Library in Cleveland. The anchors in my life are faith, family, and profession; all were critical for this project. Faith gave me the optimism and value-based framework necessary to write, especially the last chapter, and my husband, colleague, and coparent Richard provided both emotional and substantive research support. Special thanks to my father and mother, retired president and professor respectively of a historically Black college (Claflin College), who gave me the courage and confidence necessary to become something so unusual, to our world, as an urban planning professor.

Marsha would like to thank June, whose idea this book was, for asking me to share this endeavor; my institution, Virginia Tech, for approving a study leave that gave me the time to devote to the project; Neamath Taquil Douglass, for help with indexing; and the staff in the General Reference Room at the Schomberg Library of Black Culture of the Public Library of New York, who provided me with critical help in my research endeavors. In my personal life, I thank my son, Jon Ritzdorf, whose study habits are so admirable that he keeps his mother working hard to emulate them; my friends, who provide constant life support; and finally, my husband, Paul Brozovsky, for his love and patience as I single-mindedly worked on this manuscript.

CHAPTER 1

INTRODUCTION

JUNE MANNING THOMAS
MARSHA RITZDORF

A crisis mounts in urban America.[1] In central cities throughout the country, the poor struggle daily to make ends meet. Despair replaces the optimism of earlier decades, and for many young people, trapped in the new American ghetto, dangerous, nihilistic attitudes are emerging. Twenty percent of all American children, including half of African American children and 40 percent of Hispanic children, live in households below the poverty line. "The ghettos bear the accumulated weight of all the bad in the country's racial history, and they are now among the worst places to live in the world."[2] Regardless of the successes of the civil rights movement of the 1960s, and regardless of the continued growth of a Black middle class, African Americans still earn less than Whites. The past decade has witnessed staggering growth in the economic, social, and residential isolation of far too many African American families.

If urban planning is to support the equitable distribution of public goods and services, it must recognize and address the dismal conditions of millions of Americans who are poor or people of color. The primary

1

focus of contemporary planners and planning students should be on finding and advocating solutions that help eliminate the problems of today's cities. Any meaningful solution will need to be grounded in a thorough understanding of the race, gender, and class inequalities of American life.

As planners wrestle with the problems facing today's central cities, it is important to draw on all the intellectual tools possible to understand how this situation came to be and how it affects planning efforts. Those who work in conditions of suburban prosperity rather than central-city decline also need such information; they must understand why such stark contrasts linger. Planning history, a field that has flourished in recent years, is an important part of planners' intellectual arsenal. Yet as presently constituted, the field of planning history often gives inadequate preparation for understanding the relationship between planning and race.

As an example, books that focus on urban planning before World War II may discuss racial zoning but give only scant attention to the connections between residential controls and racial oppression or to the processes of ghetto formation.[3] Historical reviews of the planning profession are likely to leave out the subject of race or to mention it sporadically. Although those works that treat public housing or urban renewal often consider race as a matter of focus or necessity, authors may not make clear the relationship between race and local policy in other fields of action. Or they may ignore the complex interrelationships of race, gender, and class. For example, while public housing tenants in most cities are overwhelmingly African American households headed by single females with children, it is the exceptional book or article that acknowledges this situation and directly addresses the specific needs of citizens with that profile.

Some planning historians address the subject of race but then mishandle it. One recent book acknowledged the importance of race in public housing and urban renewal policies and, to its credit, identified the existence of an "underclass" as an issue of major importance to today's planners. Yet the author overemphasized the fragmentation of the African American family, basing his analysis on a culturally myopic vision of what defines a family. He offered little explanation of the government policies, economic restructuring, and racial and income segregation that helped create the poor in today's urban areas. This approach left readers with a disturbingly skewed picture of the causes of contemporary urban poverty and with no explanation of the relationship between racial injustice and current urban planning practice.[4]

It is time to expand the use of African American or Black urban history as a richer way of understanding the rise and development of U.S. cities in general and of U.S. urban planning in particular. The relationship between African American history and planning history is much closer than is traditionally acknowledged. In fact, one cannot fully understand the history of U.S. urban planning without understanding something about the Black urban experience, including the initiatives of community organizations, activists, planners, and politicians on behalf of their own communities.

One of the most significant and dramatic stories in the history of twentieth-century U.S. cities has been the growth and evolution of the African American population. In the early 1900s, the African American population was simply one of many ethnic and racial groups living in U.S. cities. By the 1950s, massive immigrations from the rural South to the urban North had changed the complexion of cities. By the 1990s, successive waves of in-migration by rural African Americans and out-migration by mobile Whites had created several predominantly Black cities.

African Americans became so visible in many central cities that some scholars defined their predominance and spatial isolation as indications of city decline. Indeed, throughout the twentieth century, racial prejudice shaped the lives of Blacks as surely as it shaped metropolitan areas. Long after officially sanctioned racial prejudice subsided, racial oppression and inequality lingered. Poverty grew more concentrated, and the quality of social life unraveled. Physical deterioration became the norm.

The twentieth century also witnessed the evolution of professions that were dedicated to improving urban life and reducing urban decline. Prominent among these was urban planning. Branching off from the municipal reform movement, and away from the social work and housing reform movements, urban planning aimed to create well-planned, orderly cities that allowed people to live free of slums, blight, and physical disorder. As the planning profession evolved, its practitioners attacked various maladies affecting urban areas. They joined efforts to remedy social problems, and they created initiatives designed to redevelop specific areas, such as central business districts. From the early part of the century, when planning focused on creating land use controls and regulating growth, to the end, when planners did these things plus many more, the profession's stated goal was to improve the experience of urban life for all residents. However, the reality was often far different.

Throughout the twentieth century, the community of urban African Americans connected with the community of urban planning professionals. At times those connections were sources of conflict and oppression, at other times sources of reform and cooperation. Planning tools were and are often used for the purpose of racial segregation. Examples are exclusionary zoning laws and separatist public housing programs. Urban renewal clearance projects that bulldozed Black communities into oblivion could also be classified as oppressive. But these were not the only interactions between the Black urban population and the profession. During the 1960s, collective public guilt generated basic changes in urban planning professionals as well as in national policies. Some planners—whose ranks gradually became more diversified racially—dedicated their lives to fighting for the rights of the poor and distressed. Such dedication took the form of "social" or "advocacy" planning, neighborhood planning, or equity planning.

The precise nature of this dualistic relationship of conflict versus cooperation needs further clarification. Few historians of urban African Americans give full and impartial treatment to the role of urban planning. Few historians of U.S. urban planning acknowledge the full influence of race and racial injustice on the profession. Contributions made by African American women to urban planning efforts are underappreciated.

In general, what is needed is an overview of the critical linkages between the urban planning profession and the nation's most visible racial minority. Race and racial injustice influence all efforts to improve urban society. Urban planning, an active profession, purports to help improve civic life in metropolitan areas. It cannot do so unless its practitioners more clearly understand the historical connections between this people and this field. The purposes of this book are to help clarify these historical connections and to suggest means by which cooperation and justice (as opposed to conflict and oppression) may be increased.

We have organized authors' contributions into four parts: zoning and real estate, planning and public policy, African American initiatives and responses, and planning education. We have also included, in a fifth part, a few brief excerpts from documents that are important primary sources on urban planning and race.

ZONING AND REAL ESTATE

The period between the late 1800s and World War II was characterized by major immigration and migration in the United States. In the 1880s alone, more than 5.5 million people moved from other countries to the United States, and millions more followed. But movement within the country was also massive. The demand for workers during and after World War I triggered the migration of 1.5 million African Americans from the South to Northern industrial cities.

Far more came between the early 1940s and the late 1960s, when over five million African Americans migrated from the South to the North. In sheer numbers it was the largest and most rapid mass internal movement of a people not caused by the immediate threat of extinction or starvation.[5] The migration changed the United States from a country where race could be viewed as an isolated, Southern issue to a country where race relations affected the life of people in almost every city and suburb.

By the 1920s, anti-immigrant and anti-Black sentiment was rampant among American Whites. Relying on erroneous but pervasive pseudo-scientific theories about the genetic basis of moral and intellectual characteristics, some scholars went so far as to rank the acceptability of America's ancestral groups. They placed "Anglo-Saxons and their Nordic cousins from Northern and Western Europe at the top and Eastern Europeans, Orientals and Blacks at the bottom."[6]

The major connection of these events with urban planning occurs with the rise of residential controls. One of the most important planning-related developments during the period between the world wars was the establishment of zoning as a tool for social segregation. The major actors in the landmark zoning case of *Village of Euclid v. Ambler Realty Corporation* (1926) included men who held strongly exclusionary views of African Americans and European immigrants.[7]

Another popular device was the racially restrictive covenant, a private contract limiting home sales or rentals to Blacks or Jews. The U.S. Supreme Court upheld restrictive covenants in *Corrigan v. Buckley*, which it decided in 1926, the same year that it upheld zoning in *Euclid*.[8] These two devices—zoning and racially restrictive covenants—effectively protected socioeconomic segregation and limited the mobility of people of color.

Although *Buchanan v. Warley* apparently declared racial zoning unconstitutional in 1917, unchallenged discriminatory ordinances allowed residential segregation to become entrenched.[9] Particularly in Southern cities, zoning became a powerful tool for maintaining class and racial segregation. It was not until 1948, in *Shelley v. Kramer*, that the U.S. Supreme Court declared it unconstitutional to use the courts to enforce racially restrictive covenants. The Court did not make such covenants illegal, however.[10]

Chapter 2 is basic to a historical understanding of such events. In "The Racial Origins of Zoning in American Cities," Christopher Silver examines the roots of zoning and its relationship to race by looking at the land use decisions made in Southern cities in the early twentieth century. He shows that discriminatory social objectives were central to the early planning movement and that people used these objectives to build a base for exclusionary zoning throughout the United States.

In Chapter 3, "Locked Out of Paradise: Contemporary Exclusionary Zoning, the Supreme Court and African Americans: 1970 to the Present," Marsha Ritzdorf updates the story. In the period between their 1926 and 1974 approvals of zoning as a legal use of the police power, the U.S. Supreme Court was virtually silent on issues related to zoning. This chapter reviews their reentry into zoning litigation in the 1970s and provides a social analysis of the contemporary Court's approval of certain zoning tools. These tools, in the view of most scholars, clearly aggravate racial discrimination and affect planning regulation to this day.

Raymond Mohl's Chapter 4, "The Second Ghetto and the 'Infiltration Theory' in Urban Real Estate, 1940-1960," examines a significant twenty-year period in the great rural-to-urban migration of African Americans. He begins with a description of the migration and the resulting urban violence and then outlines the real estate industry's reactionary response to the new arrivals, as revealed through an examination of real estate journals and trade magazines of the time. He shows how the real estate industry used dominant or "infiltration theory" to support and perpetrate residential segregation.

Chapter 5, by Ritzdorf, "Family Values, Municipal Zoning, and African American Family Life," concludes this section. Ritzdorf criticizes the cultural construction of the White, traditional nuclear family as the only acceptable normative lifestyle. She then explains how use of this ethnocentric construct has punished African American families by generating housing, planning, and zoning policies that limit the access and rights of those who live, by choice or by chance, in nontraditional family configurations. Such configurations are proportionately higher in the African American community.

PLANNING AND PUBLIC POLICY

The period after World War II saw two simultaneous processes: (1) the movement of the White middle and working classes to the suburbs, a movement spurred by the return of World War II veterans and the assistance of home mortgage insurance programs, and (2) the consolidation of ghetto boundaries. It is for this era that we have the best documentation concerning the relationship between African American urban life and planning decisions. As several scholars have demonstrated, political leaders' desire to shape Black residence patterns profoundly influenced public housing and urban renewal policies. Just as urban migration of rural Blacks and other ethnic minorities was the demographic motivation for racially exclusionary zoning and restrictive covenants during the period between the world wars, the need to contain Blacks in restricted sections of cities influenced public policy decisions after World War II.

The movement to the suburbs by the White middle and working classes, which one author calls a true "metropolitan revolution," clearly established decentralization as the dominant urban pattern for the following decades.[11] This decentralization, however, was exclusionary. For example, Levittown, New York, a well-known suburban community that set the pattern for numerous others, housed 82,000 residents in 1960, not one of whom was African American.[12] Although White families found new opportunities opening up in freshly constructed suburbs, African American families experienced disproportionate overcrowding and limited mobility within the central cities left behind.

A series of federal policies set the stage for these conditions. Urban renewal was one of the most invidious. Often called "Negro removal" by critics, it provides countless examples of the interconnection of racial change with local policy. Urban renewal systematically destroyed many African American communities and businesses and, for most of its history, failed to safeguard the rights and well-being of those forcibly relocated from those homes and businesses. That clearance for urban renewal worked in conjunction with clearance for highway construction only made matters worse. Backed by the federal government, cities simultaneously cleared out slums and displaced racial minorities from prime locations for redevelopment and highway construction. These policies shaped and defined the Black ghetto.

The 1960s, the era of civil rebellion, brought several important changes. The widespread civil disorders, which were volatile but predictable responses to long-standing racial oppression, forced significant alterations in federal policies. President Lyndon Johnson, attempting to build a "Great Society," initiated new programs that focused on eliminating poverty and empowering low-income communities. With the War on Poverty's community action agencies, citizens gained the power to supervise community improvement directly. Under Model Cities, local citizen governing boards also helped direct local redevelopment and made their own contributions to the redefinition of urban planning.

Well-known planning practitioners began to question the assumptions of traditional land use and redevelopment planning as well as the racial bias inherent in the profession. Proponents for advocacy planning suggested that the appropriate response to inner-city conditions was for planners to stop trying to represent the public interest—an impossible task, leading planners to represent the status quo—and to work instead to help empower disenfranchised groups. Another response was for planners to develop "suburban action" programs promoting racial and income integration. Paul Davidoff, premier advocate planner and champion of suburban integration, urged planners to champion nonexclusionary fair housing laws, low and moderate housing, and progressive zoning and subdivision requirements.[13]

What has happened to these movements? As Krumholz demonstrates in his chapter, in many ways advocacy planning is alive and well, in the form of contemporary "equity planning." But as Rabin shows in his chapter, previous attempts to use federal policy to address many of these concerns have stumbled, and racial isolation has persisted.

Rabin focuses on employment and civil rights policies and decisions, but other examples exist as well. The Housing and Community Development Act of 1974 killed the oppressive urban renewal program, but it also brought the promising Model Cities experiment to a halt. With the 1974 act, which created Community Development Block Grants (CDBGs), the federal government withdrew from high-profile attempts to target funds to distressed central-city efforts, defined and guided by local citizens. Instead, in city after city, citizens who had just begun to exercise some control over the redevelopment of their neighborhoods experienced the shock of governmental withdrawal. Although in later years the CDBG program somewhat improved on its record of participation, in general the program placed decision making in the hands of city government and dispersed national funding via a formula that spread increasingly scarce redevelopment funds to populous suburbs as well as to a wide range of cities.

Previous efforts to mesh social, economic, and physical development strategies, a mixture allowed under Model Cities, succumbed under the pervasive "bricks and mortar" orientation of the CDBG program. Any illusions that inner-city residents might have had that a benign federal government would "gild" their ghetto died quickly with the unstable funding, unpredictable longevity, and strong downtown focus that characterized urban-related programs such as action grants and economic development assistance funds in the 1970s, 1980s, and early 1990s. The mid-1990s brought promising federal program initiatives, such as Empowerment Zones/Enterprise Communities. But by that time African American families, even those in suburbia, remained highly segregated. They earned less money than others per capita and per family, and experienced much narrower options of residence than did other Americans.

Chapters in this part of the book touch on some of these issues. Yale Rabin begins with "The Persistence of Racial Isolation: The Role of Government Action and Inaction" in Chapter 6. Rabin reviews thirty years of studies, government programs, and court decisions, particularly concerning transportation, employment, and civil rights, which have either failed to question or aggravated racial isolation. He argues that current urban conditions are the effect, not the cause, of isolation and discrimination. His spirited essay charges that cynical politicians have embraced a philosophy that perpetuates injustice and that blames the victims of federal and local policies.

Norman Krumholz further documents the poor status of inner-city minorities in Chapter 7, "Urban Planning, Equity Planning, and Racial Justice." While reminding us of the conscious, historic tradition among some planners to devise and implement redistributive policies, he juxtaposes this against the small numbers of people of color engaged in planning as a profession, and the small number of planners, overall, who embrace equity ideals. Yet he enthusiastically describes the importance of the equity planning tradition and reviews for the reader several successful equity planning projects around the nation, holding them up as models for others to follow.

The next chapter, Chapter 8, makes the heretofore fairly broad juxtaposition of local policy, planning, and race come into focus through a case study of Gary, Indiana. Robert Catlin's "Gary, Indiana: Planning, Race and Ethnicity" is a cautionary tale that examines Gary from 1900 to the present. As Catlin explains, the dynamics of the interaction of race and planning strongly influenced Gary's rise and decline as a Midwestern city. He argues that institutional racism contributed to a political climate that hastened Gary's economic and physical decline.

June Manning Thomas concludes the part in Chapter 9, "Model Cities Revisited: Issues of Race and Empowerment." There, she looks forward at contemporary urban initiatives by first looking backward at Model Cities. Using a methodology that demonstrates the potential for inferential qualitative research in this field, she uses the framework of African American community concerns to reassess Model Cities in Cleveland and Detroit. Model Cities had several decided benefits, it seems, and the experience provides relevant lessons for contemporary urban programs, especially Empowerment Zones/Enterprise Communities.

AFRICAN AMERICAN
INITIATIVES AND RESPONSES

Unfortunately, much of the writing about the relationship between the African American community and urban planning has focused on victimization. Of course, victimization, injustice, and oppression are important parts of the story. But throughout the twentieth century, African Americans have refused to be passive actors in this process. They documented their situation, built indigenous institutions, and undertook initiatives designed to improve community life. Scholars such as W. E. B. DuBois carried out path-breaking research, and organizations such as the National Urban League and the National Association of Colored Women made major contributions—which while documented in other ways are undocumented in the annals of planning history—to planning efforts in their own communities.

Early in the century, African American women often focused on the civic improvement of their communities. While they, like White women, had no legal or voting rights in the public world of politics, they were very active. Yet they, like their African American brothers, are invisible from the records of their time that planning historians commonly consult. For example, *The American City*, a periodical that began publication in 1909, was "the" source of information about urban issues, problems, and projects throughout the early part of this century. Between 1909 and 1920, only one article in any way related to African Americans, and it concerned the creation of a segregated low-income housing project. In 1912, an entire issue reported on women's contributions to civic improvement, but it reported only on White women's organizations. Future work will need to look at the contributions of women who participated in projects linked to traditional urban planning, such as housing, parks, land projects, and sanitation, or who made a place for themselves in male-dominated organizations such as the Urban League.

The Urban League exemplified African American leadership and response to planning throughout much of the twentieth century. During the years of migration local chapters actively sponsored day camps, food drives, employment programs, and numerous other activities. In the 1950s, these chapters were often leaders in the efforts to document the initial abuses of the urban renewal program. The Chicago branch's 1968 report, *The Racial Aspects of Urban Planning: Critique on the Comprehensive Plan of the City of Chicago*, portions of which we have reproduced at the end of this book, clearly identified the role of institutional racism in the planning process and offered proposals for change. As they noted, "Abstract statements about the goal of equality, while welcomed, are no substitute for technical work dealing with the realities of racism."[14]

By the 1970s, African American communities began to realize that environmental problems in their communities were related to discriminatory exposure to both toxic substances and unwanted land uses. Lead poisoning, especially from exposure to lead-based paint in substandard urban housing, was an issue of social justice that demanded their attention. The combined efforts of inner-city activists and a small group of physicians/scientists ultimately forced the issue onto the public agenda. A Philadelphia coalition brought a lawsuit against the federal Department of Housing and Urban Development (HUD) to ensure that HUD property was inspected, and if necessary, cleaned of all offending lead.[15] Over the next two decades, groups identified myriad other urban environmental issues and added environmental justice to their civil rights agendas.[16]

A range of other kinds of African American self-help efforts have persisted in recent years, particularly community development. Rather than wallow helplessly in defeatism, Black politicians, faith-based groups, and community-based organizations in some cities have carried out remarkable, heroic efforts to preserve and improve their communities. These initiatives addressed a myriad of issues, including but not limited to redevelopment, housing rehabilitation, redlining by financial and insurance institutions, commercial development, and social improvement programs for youth and families.

It is because of this proud tradition of African American activism that we are particularly pleased to include the four chapters in this part of the book. We present them in approximate chronological order. In Chapter 10, "Charlotta A. Bass, the *California Eagle,* and Black Settlement in Los Angeles," Jacqueline Leavitt introduces us to an early, feisty advocate for racial justice. Editor of the African American *California Eagle* for nearly forty years, Bass fought residential segregation efforts in Los Angeles. She focused particularly on private contracts, known as racially restrictive covenants, that prohibited the sale and ownership of property to African American residents of Los Angeles and other U.S. cities and suburbs.

Next, Sigmund Shipp presents a case study of urban renewal and the African American community in Chapter 11, "Winning Some Battles But Losing the War? Blacks and Urban Renewal in Greensboro, N.C., 1953-1965." He examines the origin and development of the Cumberland Project, Greensboro's first urban renewal project. His chapter challenges

the traditional interpretation that African Americans were noncoopera-
tive victims of urban renewal. He shows that middle-class African
Americans and their institutions supported Greensboro's redevelopment project. Although redevelopment devastated many African American businesses, he suggests that the displacement that occurred may
have been balanced by improved housing available in other sections of
the city.

Charles Connerly and Bobby Wilson trace years of a proud African
American community planning tradition and its influence on contemporary planning in a Southern city in Chapter 12, "The Roots and Origins
of African American Planning in Birmingham, Alabama." By sketching
the historic role of civic leagues and their role as civic improvement
organizations in Birmingham's neighborhoods, they show how the African American community was able to reshape and redefine Birmingham's citizen participation plan, which is nationally recognized for the
citizen involvement it generated in the planning and development processes of the city.

Chapter 13, "Urban Environmentalism and Race," by Robert Collin
and Robin Morris Collin, ends this part by looking at the urban environmental movement from before the 1970s to the present. The authors
document the rise of environmental concerns as a focus of African
American activism during this time period and summarize the importance of conferences, reports, and law cases that provide the core materials that community activists now rely on. As they note, the importance
of urban environmentalism extends beyond issues of minority inclusion.
The urban environmental or environmental justice movement has redefined traditional environmentalism in creative and important ways,
providing an impressive set of alternative strategies for environmental
problem solving.

PLANNING EDUCATION

An important area of concern regarding the relationship of urban
planning with race is planning education. Many of today's urban and
regional planners receive their first exposure to the profession in undergraduate or graduate programs in urban planning. These programs also
house scholars, faculty members, and advanced graduate students who

think and write about issues related to the metropolis, local and federal policy, and the planning profession. If we are to reshape how urban policymakers and professionals think about race and planning, both from a historical and a contemporary perspective, we must influence these university programs.

The classroom is a formative experience for most planners, offering a context for promoting a view of planning and development that is more sensitive to the ways that race, gender, and class influence the lives of citizens and frame the discourse of contemporary planning policy. Planning students must understand the racial context of cities so that they can appreciate how planners in the past have wrestled with issues that still haunt us. In the process they may learn skills useful for their own survival in the realities of contemporary city planning. We waste valuable theories when we do not use past experiences to illuminate present problems.

Planning students and scholars have much to learn from racially conscious studies of planning history. They can begin to understand how deeply the roots of racial segregation reach into U.S. urban history and how strongly implicated zoning and residential controls have been in society's efforts to keep races separate. This could help them treat tools such as zoning with appropriate caution and learn ways to use these tools in a more progressive way. The area of postwar public policy also offers important lessons. Racially conscious studies of urban policy history can help explain why the racial divisions of cities remain so strong, why public housing units are so segregated and distressed, and why distrust of redevelopment lingers in the Black community.

With additional discussion of the interaction of gender, race, and poverty, planners can learn to explicitly address the needs of African American women and their children as they develop housing and social service programs. Without these sensitivities, new programs are likely to contain the same flawed assumptions and solutions that haunt us from past efforts at community redevelopment and citizen involvement.

Another important issue is the need to expand the voice of racial minorities within the planning profession and planning academia. Many of the concerns listed earlier could be addressed in part by increasing diversity within the profession. By recruiting as planners more people who have grown up in inner-city African American communities and who therefore already know about many of the challenges facing those communities, we could greatly increase the ability of the profession to develop new, supportive strategies for inner-city development. Conversely, with greater diversity, the profession could more easily offer urban society an array of existing planning tools that could be very useful.

The last two chapters in this book address this issue of planning education in two different but important ways. Siddhartha Sen reports on "The Status of Planning Education at Historically Black Colleges and Universities: The Case of Morgan State University," in Chapter 14. This is simultaneously a disturbing and an uplifting chapter. It is disturbing in its frank explanation of the failing struggle of urban planning programs to arise and survive at historically Black colleges and universities (HBCUs). It is uplifting in its description of the important role HBCUs have played in the training of African American planners and scholars. The chapter pays special attention to the Morgan State program, with its strong commitment to improving the Baltimore central city and increasing the number of Black planners.

In Chapter 15, "Coming Together: Unified Diversity for Social Action," June Manning Thomas lays out a visionary plan for diversifying planning education in order to improve academia and the profession. Drawing on several sources—the literature on multicultural education; the sparse, but relevant, literature within planning education; and her life experiences as an African American planning faculty member—she suggests a set of ideas that could be incorporated into academic planning programs and curricula. These would provide a diverse and socially progressive framework of effective study for all students but particularly for those whose needs have received insufficient attention in the past: African Americans, other racial minorities, and women.

PLANNING-RELATED DOCUMENTS

The last part of this book includes excerpts from several planning-related court cases, government documents, and reports, so that readers may glimpse for themselves the primary materials that form the historical basis for this book. We have included these materials in chronological order. We begin with excerpts from a 1926 U.S. Supreme Court case that legitimized zoning but established the supposed inferiority of multifamily housing, referred to as "parasites": *Village of Euclid v. Ambler Realty Corporation*. Excerpts from the Federal Housing Administration's 1938 *Underwriting Manual* demonstrate how completely government workers with a great deal of power to shape suburban development believed in restrictive covenants and racial segregation. Excerpts concerning relocation from a 1961 urban renewal manual demonstrate that redevelopment planners were supposed to be sensitive to the specific housing needs of non-White people, although, as the commentary notes, they often were not. The 1968 Chicago Urban League's report on Chicago's Comprehensive Plan gives an important perspective on African American opinions about urban planning, and we only regret we could not include more of this document.

Three of the excerpts date from the 1970s. The first is Justice Thurgood Marshall's dissent in the 1973 *Village of Belle Terre v. Boraas* case, in which the Court supported restricting the rights of unrelated individuals to choose their living companions and place of residence. We've also cited Justices Douglas and Brennan in their dissent to the Court's decision, in the 1974 *Warth v. Seldin* case, not to allow area low-income residents to challenge exclusionary zoning in the town of Penfield, New York. An excerpt from a 1976 commentary by scholar Milton Morris of the Joint Center for Political Studies discusses the racial implications of the Housing and Community Development Act of 1974.

The last four documents date from the 1980s and 1990s. Selections from the 1988 U.S. Fair Housing Act explain the scope of the protection that the Act offers against discrimination, as well as certain exceptions that apply to rental housing. A copy of portions of a recent version of the American Institute of Certified Planners Code of Ethics is a fitting excerpt to accompany several of the chapters (e.g., Silver's and Krumholz's), since it reconfirms the importance of the urban planning profession committing to a new era of justice, equity, and fair play. President Clinton's 1994 executive order on environmental justice (No. 12898) launched an important effort to establish federal policies responsive to the need to alleviate environmental hazards affecting minority and low-income populations. Another important initiative during the Clinton administration was the attempt to improve community reinvestment laws; the last document summarizes recent findings that confirm the benefits of this endeavor, but suggest that problems remain.

A FINAL NOTE

As we said at the beginning, one purpose for this book is to illuminate the historical connections between urban African Americans and the urban planning field. Several of our chapters do this well, documenting the racial impacts of zoning and restrictive covenants, the insidious role of the real estate industry, the effects of federal policy neglect, and the foibles of local decision making. We also provide important information about the roles of equity planning, the Model Cities experiment, community planning, minority environmental activism, and historically Black schools. But we indicated a second purpose: to suggest means for increasing justice and cooperation between urban planners and the African American urban community.

Upon reflection, it is easier to fulfill the first purpose than the second. Documenting conflict, abuse, and injustice is easy enough; looking for cooperative solutions and realistic, problem-solving visions of justice is not. But we are pleased with this humble effort, which (from our not unbiased view) does begin to suggest a few feasible means of cooperation and solution building for social justice.

Note, for example, the implications of our chapters on zoning. Rectifying solutions are readily apparent. We must first recognize the continuing problems caused by use of the tool that planners so heavily rely on—zoning. At some point we will need national judicial leadership to resolve many lingering issues, but in the meantime, planners can educate themselves and encourage their local politicians and commissions to resist the overt or covert tendency to develop exclusionary communities. The chapters suggest that redefining regulations and unconscious assumptions could go a long way toward increasing social justice.

In the planning and policy section, all of the chapters illustrate the importance of taking steps to build bridges between fragmented races and classes. While equity planning is the clearest positive vision of what this would mean, the other chapters suggest that we change federal, state, and local policies to address dilemmas of disunity, noncooperation, and oppression. These chapters include several recommendations to help today's policymakers avoid contemporary urban programs that are ineffective or insensitive to urban constituents.

The last two chapters of the section on African American initiatives offer particularly good ideas about increasing contemporary cooperation and effective social justice. From the example of Birmingham, we learn that encouraging indigenous, African American community planning, and meshing that planning with municipal planning structures, can bring forth stellar benefits. As urban-environmental issues become more significant to the majority of Americans, paying attention to the tactics of savvy minority urban environmentalists can help us learn how to resolve the stickiest environmental problems facing us today.

Finally, the chapters on planning education contain two key messages. One is that we should better encourage and support the precious few planning programs remaining at historically Black universities and perhaps start more. The other is that we should remake our planning schools so that they become climates of cooperation and inclusion. Only by changing the environments where planning education occurs can we, by example, provide meaningful solutions to the problems planners face in fulfilling their ethical obligations to others in a rapidly diversifying world.

NOTES

1. Portions of this introduction originally appeared in June Manning Thomas, "Planning History and the Black Urban Experience: Linkages and Contemporary Implications," *Journal of Planning Education and Research* 14.1 (1994): 1-11. Permission obtained for reuse.

2. Nicholas Lemann, *The Promised Land: The Great Black Migration and How it Changed America* (New York: Knopf, 1991), 353.

3. Charles Harr and Jerold Kayden, eds., *Zoning and the American Dream: Promises Still to Keep* (Chicago: Planners Press, 1989); Mel Scott, *American City Planning Since 1890* (Berkeley, CA: University of California Press, 1971); Christine Boyer, *Dreaming the Rational City: The Myth of American City Planning* (Cambridge, MA: MIT Press, 1986).

4. Peter Hall, *Cities of Tomorrow: An Intellectual History of Urban Planning and Design in the Twentieth Century* (Oxford: Basil Blackwell, 1989).

5. Lemann, *Promised Land,* 6.

6. Arthur Mann, "From Immigration to Acculturation," in *Making the Society and Culture of the United States*, ed. Luther S. Luedtke (Chapel Hill, NC: University of North Carolina Press, 1992), 75.

7. William Randle, "Professors, Reformers, Bureaucrats," in Haar and Kayden, *Zoning and the American Dream*; Yale Rabin, "Expulsive Zoning: The Inequitable Legacy of Euclid," in ibid.; *Village of Euclid v. Ambler Reality Corporation*, 272 U.S. 365 (1926), 465 S. Ct. 114.

8. *Michigan v. Buckley* 299 F. 899 (1924); 271 U.S. 323 (1926).

9. Randle, "Professors, Reformers, Bureaucrats."

10. *Shelley v. Kramer* 334 U.S. 1 (1948).

11. Kenneth Fox, *Metropolitan America: Urban Life and Urban Policy in the United States, 1940-1980* (New Brunswick, NJ: Rutgers University Press, 1990).

12. Kenneth Jackson, *The Crabgrass Frontier: The Suburbanization of the United States* (New York: Oxford University Press, 1985), 240.

13. Paul Davidoff, "Advocacy and Pluralism in Planning," *Journal of the American Institute of Planners* 31: 331-38; idem, "A Rebuilt Ghetto Does Not a Model City Make," *Planning 1967* (Chicago: American Society of Planning Officials, 1967).

14. Urban League of Chicago, *The Racial Aspects of Urban Planning: Critique on the Comprehensive Plan of the City of Chicago* (Chicago: Urban League of Chicago,1968), 11.

15. Robert Gottlieb, *Forcing the Spring: The Transformation of the American Environmental Movement* (Washington, DC: Island Press, 1993), 247.

16. Gottlieb, *Forcing the Spring*; Robert Bullard, "Environmentalism and the Politics of Equity: Emergent Trends in the Black Community," *MidAmerican Review of Sociology* 12.2 (1987): 21-37.

PART ONE

ZONING AND REAL ESTATE

See Part 5.A, B, E, F, and H,
for documents related to
zoning and real estate.

THE RACIAL ORIGINS OF
ZONING IN AMERICAN CITIES

CHRISTOPHER SILVER

The introduction of zoning in the early 1900s launched a revolution in American land use regulation and planning. Beginning with height regulations in Washington, D.C., in 1899, efforts to control the type and intensity of land use spread to many cities. In 1908, Los Angeles adopted the nation's first citywide "use" zoning ordinance to protect its expanding residential areas from industrial nuisances. Over the next two decades, state legislatures nationwide granted to cities the power "to regulate the height, area, location, and use of buildings in any designated part or parts of their corporation limits." The U.S. Supreme Court's sanction of this exercise of a city's police power over land use came first in *Hadacheck v. Sebastian* (1915), which involved the Los Angeles ordinance, and culminated in the definitive *Village of Euclid v. Ambler Realty Corporation* case in 1926 (see planning-related documents, Part 5.A).[1]

The tendency of planning historians to focus on land use regulations principally as a way to shape the built environment and to stabilize land values obscures equally important (and less publicized) social objectives in America's early planning movement. In *Zoning and the American*

Dream, Charles Haar points to the diverse interests that coalesced in the early 1900s to create the "remarkable socio-legislative phenomenon" of zoning. Haar contends that a "ragtag grouping of idealists and special interest groups of the most diverse origins" looked to zoning as a tool for social reform as well as land use control.[2] These social reformers believed that zoning offered a way not only to exclude incompatible uses from residential areas but also to slow the spread of slums into better neighborhoods. Reformer/planner Benjamin Marsh championed zoning in the early 1900s in an effort to combat urban congestion and thereby improve the quality of working-class neighborhoods. Despite the obvious social implications of early zoning initiatives, however, the noblest intention of reformers like Marsh soon gave way to political pressures from those less inclined toward broad civic improvement. "What began as a means of improving the blighted physical environment in which people lived and worked," writes Yale Rabin, became "a mechanism for protecting property values and excluding the undesirables." The two interest groups that were regarded as the undesirables were immigrants and African Americans.[3]

Rabin's study emphasizes the "social origins" of zoning and planning in the United States. He notes, as have other scholars, that Southern cities in the early twentieth century used zoning to enforce the newly created system of racial segregation. "While northern Progressives were enacting zoning as a mechanism for protecting and enhancing property values," Rabin observes, "southern Progressives were testing its effectiveness as a means of enforcing racial segregation."[4] Baltimore enacted the first racial zoning ordinance in 1910; within several years the practice was widespread in the region. The racial zoning movement received a sharp reversal in 1917, when the U.S. Supreme Court declared a Louisville, Kentucky racial zoning ordinance unconstitutional. Despite the Court's ruling in *Buchanan v. Warley,* Southern cities persisted in seeking a legally defensible way to use zoning to control Black residential change. In the place of race zoning per se, Rabin contends, many cities turned to "expulsive zoning," which permitted "the intrusion into Black neighborhoods of disruptive incompatible uses that have diminished the quality and undermined the stability of those neighborhoods." The concept of "expulsive zoning" helps to explain how American cities made the transition from racial zoning to recent zoning that has a decidedly discriminatory impact on Black neighborhoods.[5]

RACIAL OBJECTIVES OF ZONING

The purpose of this chapter is to examine the "racial" roots of the U.S. zoning movement by examining Southern cities both prior to and immediately following *Buchanan v. Warley*. The racial zoning movement in the urban South demonstrates clearly how certain social objectives were central to the early planning movement. While scholars have examined the racial zoning movement leading up to *Buchanan v. Warley*,[6] they have given relatively little attention to important racial zoning initiatives after 1917. It is in this post-1917 period especially that cities hired prominent planning professionals to fashion legally defensible racial zoning plans.[7] Throughout the early 1900s, and well beyond 1917, racial zoning and its objectives remained a mainstay of many American planners. Racial zoning was not just a manifestation of the backward South out of touch with the mainstream of urban reform. Although the South invented and made wide use of racial zoning, the region relied on Northern planning consultants to devise legally defensible ways to segregate Black residential areas.

Racial zoning practices also transcended the South. Select Northern and Western cities, especially those where the Black population increased rapidly, also experimented with racial zoning. The nation's planning movement, not just its Southern branch, regarded land use controls as an effective social control mechanism for Blacks and other "undesirables." According to H. L. Pollard, a prominent Los Angeles land use attorney, "racial hatred played no small part in bringing to the front some of the early districting ordinances that were sustained by the United States Supreme Court, thus giving us our first important zoning decisions."[8] Chicago, too, was a bastion of racial zoning enthusiasts.[9] Despite evidence that the racial zoning movement was national in scope, it initially concentrated in Southern cities owing to the relative size of the Black community (which ranged between 30 and 50 percent of the population in many places), and it then spread northward and later westward in response to the migration of Southern Blacks.[10]

It is also important to note that Southern cities experimented with racial zoning and comprehensive zoning in tandem with, and not merely in the wake of, land use regulation efforts elsewhere. The 1908 ordinance of Richmond, Virginia, to regulate the height and arrangement of buildings, which was upheld by the Virginia Supreme Court of Appeals in

1910, "was used by proponents of zoning in New York City as a precedent for persuading the city and state legislatures to act favorably on
their recommendation," and led to the landmark 1916 zoning ordinance.
Also on the basis of the 1910 decision, Richmond drafted and enacted a
racial zoning scheme early in 1911.[11] Racial zoning in Southern cities was
as much a foundation for overall land use regulations as were regulation
of the garment industry in New York City or encroaching industrial uses
in Los Angeles.[12]

The experience in Southern cities suggests that zoning, land use
regulations, and comprehensive planning proved to be effective tools to
reshape the urban social landscape. Racial zoning persistently failed to
withstand legal challenges. Nevertheless, planning that regulated urban
development through implementation of master plans and capital improvement programs, as well as through a more subtle sort of "racially
informed zoning," helped to create the racially bifurcated social geography of most contemporary American cities. In Southern cities, racial
concerns infused a wide array of public initiatives beyond zoning, which
explains why urban planning represented such an important component
of Jim Crowism in the region. Existing historical scholarship fails to give
sufficient attention to the way that planning, in concert with legal prohibitions against racial intermingling, influenced the social development
of the New South. At the same time, the social impact of zoning and
planning in the urban South has been obscured by the mythology of
reform and progress that surrounds the early planning movement in the
United States.[13]

Of course, other factors such as income, age and type of housing, real
estate practices, and culture contributed to the highly segregated residential patterns of contemporary Southern cities. Yet the racial zoning
movement launched what became a comprehensive set of public policies
to contain Black residential expansion. Despite the short legal life of
racial zoning, it continued to shadow public initiatives in community
development as late as the 1960s. In Atlanta, for example, public officials
went to great lengths to prove that their efforts to guide the expansion
of the Black community in the 1950s were not illegal racial zoning, even
though their brand of neighborhood planning effectively defined Black
and White areas. By the time Atlanta's leaders ceased to support regulation of Black neighborhood change, the city was almost completely
divided spatially between two separate worlds, one Black and one

White. In this sense, the racial zoning movement is not just an historical aberration of the pre-civil rights era but a central feature of American planning history throughout the twentieth century.[14]

EARLY RACIAL ZONING LAWS

The first comprehensive racial zoning ordinance in the United States appeared in the quasi-Southern metropolis of Baltimore in December 1910, although several California cities had for decades employed the "police power" to control the spread of Chinese laundries outside Chinese neighborhoods. Local attorney Milton Dashiel fashioned Baltimore's racial zoning plan immediately following the momentous decision of another attorney, George W. F. McMechen, to move into the fashionable Eutaw Place. With the support of Councilman Samuel L. West, Dashiel's plan to contain Black residents worked its way slowly, but methodically, through both branches of city council, despite immediate protests from Black residents.[15]

Mayor J. Barry Mahool, a nationally recognized member of the "social justice" wing of the Progressive movement, gave unequivocal support to the city's pioneering racial zoning ordinance and signed it into law on December 20, 1910. Like many reformers in Baltimore, Mahool subscribed to the position that "Blacks should be quarantined in isolated slums in order to reduce the incidents of civil disturbance, to prevent the spread of communicable disease into the nearby White neighborhoods, and to protect property values among the White majority."[16]

Passage of the Baltimore ordinance, which came years before New York City's Fifth Avenue garment retailers even began to organize their pioneering zoning initiative, unleashed a flood of similar laws in Southern cities. Over the next few years, several Virginia cities, including Richmond, Norfolk, Portsmouth, Roanoke, and the town of Ashland, enacted modified versions of the Baltimore plan. Atlanta, Georgia; Greenville, South Carolina; Asheville and Winston-Salem, North Carolina; Birmingham, Alabama; and Madisonville and Louisville, Kentucky also embraced the idea of racial zoning.[17] Others, such as Charlotte, North Carolina; Charleston, South Carolina; Meridian, Mississippi; and New Orleans considered, but did not immediately enact, racial zoning ordinances prior to 1914.[18]

Virginia's racial zoning movement got underway in 1910 as soon as the Virginia Supreme Court of Appeals upheld the constitutionality of Richmond's 1908 act to regulate the height and arrangement of buildings.[19] Virginia's enabling legislation allowed cities to zone their entire area according to race, whereas the Baltimore plan applied only to all-White or all-Black blocks and not to mixed blocks. Richmond's 1911 ordinance, passed just twelve days after enactment of Baltimore's second racial zoning law, stipulated that "a block is White where a majority of the residents are White and colored where a majority . . . are colored."[20]

Richmond's residential segregation ordinance received the blessing of the state's highest court in *Hopkins v. City of Richmond* in 1915.[21] The *Hopkins* case became a widely cited defense of racial zoning both prior to and following the Supreme Court's landmark *Buchanan v. Warley*. The *Hopkins* case involved a White and a Black who moved into a house together in a designated "White zone" after enactment of the racial zoning ordinance. The court maintained that Richmond's ordinance did not deny property rights since the complainants moved in following passage of the law. In particular, to counter "taking" objections to racial zoning, it cited the grandfather provisions of the Richmond and Atlanta laws, which allowed property ownership and right to access that property by both races in "mixed neighborhoods." As late as 1927, ten years after the *Buchanan v. Warley* ruling, proponents of racial zoning still pointed to the *Hopkins* case, as well as to a favorable lower court decision involving the constitutionality of Atlanta's 1915 racial zoning law, as proof that at least two state courts placed racial zoning within the legal limits of the police powers of cities.[22]

The courts paid little attention to the social implications of racial zoning, however. The practice of allowing ownership of property by Whites in Black neighborhoods (and, in theory, by Blacks in White neighborhoods) fostered absentee ownership and reduced the incidence of Black home ownership. Even as the Black population of Richmond moved out of its scattered residential enclaves in the early 1900s, and thereby changed the racial composition of other neighborhoods from White to Black, out-migrating Whites tended to rent rather than sell their houses to Blacks. In the absence of new housing construction, the perpetual shortage of Black housing enabled absentee landlords to profit handsomely from neighborhood turnover. In Richmond, at least, one effect of the short-lived racial zoning law and subsequent controls over Black residential migration was a reduction in home ownership in the Black community.[23]

Both the Baltimore and Richmond racial zoning campaigns drew critical support from local housing reformers. In the case of Baltimore, middle-class reformers paid particular attention to blighted housing conditions in the predominantly Black Seventeenth Ward. A 1907 report illuminated "the horrors of the slums and the plight of the slum-dwellers" and offered various improvement strategies, such as model housing, enactment of housing codes and building regulations, and removal of alley dwellings. Although the city took no formal action on the 1907 report, interest in controlling the spread of blighted housing logically translated into support for racial zoning as Blacks crossed the color line after 1910 in search of better housing. Like their leader in the mayor's office, Baltimore's housing reformers offered no resistance to Dashiel's plan for regulating neighborhood change.[24]

Richmond's reform movement produced its own catalog of housing horrors when the Society for the Betterment of Housing Conditions published an equally graphic depiction of the city's dilapidated Black neighborhoods. Released in 1913, the Society's report made no direct reference to racial zoning as a remedial action but, instead, concentrated on housing codes, building regulations, removal of alley dwellings and, especially, the need for new model housing. The 1913 report, which appeared while the city's racial zoning law was still in force, did note the importance of determining appropriate areas for new Black residential development to eliminate the demand for substandard housing in deteriorated areas. If nothing else, its silence on the matter of racial zoning served as a tacit endorsement.[25]

A third factor, besides concern over housing blight and Black encroachment in existing White neighborhoods, explains the sudden widespread interest in racial zoning among Southern cities. One assessment of the origins of neighborhood associations in Baltimore suggests that the quest for a permanently restructured city, with neighborhoods functioning as separate "urban islands," increased the appeal of the legal sanctions afforded by zoning. At a 1911 Citywide Congress of Neighborhoods in Baltimore, attended by delegates from forty-one improvement and protective associations, participants debated the merits of improving housing through either cooperation or barriers to social interaction.

While a handful of delegates believed that "social problems" could be corrected through cooperation, especially by clearing and rebuilding "undesirable neighborhoods," the Congress officially rejected cooperation in "favor of the more 'practical' segregationist policies advocated by the city planner."[26]

In Atlanta, the objective of racial zoning was legalized separation of the city into separate racial worlds. The city's racial zoning ordinance, enacted on June 16, 1913, followed the Baltimore formula except that, like Richmond, it assigned a racial designation to every city block based on the existing majority of the residents, not just to those that were already all-White or all-Black. While Baltimore reformers were engaged in a study of blighted housing conditions in 1906, Atlantans were caught up in a violent race riot. Instigated, in part, by the "reckless anti-Negro agitation" of gubernatorial candidate Hoke Smith and his journalistic supporter, Tom Watson, the Atlanta race riot resulted in the deaths of twenty-five Blacks as White mobs assailed the city's Black residents.[27]

Ever the exponent of moderation and conciliation, Booker T. Washington rejected the view that the Atlanta riot represented a step backward in race relations. Rather, he contended, it provided an opportunity for "reconstruction." E. Franklin Frazier lent support to this view. Writing from Atlanta in 1923, he noted that the city's Black community began to organize itself following the 1906 riot. Not only was "segregation shutting our colored people from the wider community of Atlanta" but the Black community shifted from the eastside to the westside and separated from the White neighborhoods. As demographic factors propelled shifts in Black settlement, the prospect of controlling residential change through zoning gained widespread support among White Atlantans.[28]

Atlanta's racial zoning ordinance failed its initial court test in 1915, when the Georgia Supreme Court ruled that the law violated state and federal protection of "rights in property acquired previous to its enactment."[29] When Atlanta revised its ordinance to exempt residences acquired before passage of the ordinance, the Georgia high court sustained the city's racial zoning plan in 1917.[30]

The euphoria of Georgia's segregationists faded quickly, however, when the United State Supreme Court unanimously struck down a Louisville, Kentucky racial zoning ordinance later that year. In the landmark decision, *Buchanan v. Warley*, the Court ruled unanimously that the denial of the full use of property "from a feeling of race hostility" constituted inadequate grounds to uphold the Louisville racial zoning ordinance.[31] With such an unequivocal ruling from the nation's highest court, lower courts fell into line and overturned existing and subsequent racial zoning schemes. *Buchanan* did not end the racial zoning movement, however, but merely shifted it to new ground, as will be noted later.

Nonetheless, long after the 1917 decision, racial zoning proponents lamented the restrictions imposed by the *Buchanan* case. As late as 1926, in a study of the housing conditions of Blacks in several Virginia cities, Charles Knight lamented the inability of state and local governments "to keep separate Negro and White residential sections" through the use of zoning. "The results of this course," he maintained, were "to increase friction between the White race and the Black" and to exacerbate already deplorable housing conditions in existing neighborhoods.[32] In the concluding section of an analysis of various municipal zoning and segregation ordinances in 1927, George D. Hott warned that

> the commingling of the homes and places of abode of White men and Black men gives unnecessary provocation for miscegenation, race riots, lynching, and other forms of social malaise, existent when a childlike, undisciplined, inferior race is living in close contact with a people of more mature civilization.

He hoped that

> public opinion may come to preponderate so strongly in favor of sustaining municipal race segregation ordinances, drafted so as to be reasonable and not to deprive of previously acquired property, that they will ultimately be held constitutional.[33]

The *Virginia Municipal Review* contended that "a gradual and natural encroachment of the colored population into White neighborhoods" was the obvious consequence of an unregulated residential market. Given that Richmond's racial zoning ordinance fell under the authority of the *Buchanan* decision, the editor noted that the city found itself "face to face with a problem of increasing significance whose solution deserves the thought and discussion of leaders of both races."[34]

RACIAL ZONING AFTER BUCHANAN V. WARLEY

The decade following the *Buchanan* decision saw numerous efforts to fashion a legally defensible racial zoning system in Southern cities and in scattered areas outside the region. Atlanta, Indianapolis, Norfolk, Richmond, New Orleans, Winston-Salem, Dallas, Charleston, Dade County (Florida), and Birmingham, to name only the most prominent places, passed new racial zoning legislation after 1917. Many others discussed the topic seriously and looked to consultants to find a workable approach to planned apartheid.[35]

This new movement to legalize residential segregation was different in several ways, however. Most of the residential zoning laws fashioned prior to 1917 were the work of nonplanners who recognized the potential of land use regulation to achieve social objectives. After 1917, cities preferred to engage professional planners to prepare racial zoning plans and to marshal the entire planning process to create the completely separate Black community. The *Buchanan* decision undermined the use of zoning to segregate explicitly by race but not the use of the planning process in the service of apartheid. Charles Knight noted in the case of Virginia that cities employed sections "designated as Negro residential areas." Even if they did not legally enforce land use, these designations guided public and private developments. Data supplied by planners made it possible to monitor and influence land use trends based on social criteria.

In this way, racial zoning still operated in practice if not in law, reinforced by a planning process that supported the creation of a racially bifurcated society. In Knight's view, this was not necessarily detrimental to Blacks, however. He contended that planning was not an impediment to Black community development but rather an essential ingredient in the full realization of a segregated metropolis. Rather than thwarting Black social development, "zoning laws should preserve the residential character of the (Black) areas." Black neighborhoods "should be desirably located with respect to topography, industry, and convenience as the White areas," and should be large enough to accommodate future population growth. Finally, he observed, Black neighborhoods should benefit from "all necessary municipal services—paving, city water, sewers, electricity, fire and police protection," as well as sufficient parks and playgrounds and laws "to prevent housing and land crowding."[36]

The community development strategy outlined by Knight required far more than zoning laws to prevent cohabitation in neighborhoods on the basis of race. To bring about a totally separate Black community necessitated a comprehensive planning effort. This is what happened in a number of Southern cities. In the wake of the *Buchanan* decision, racial zoning gave way to the broader notion of a race-based comprehensive planning process. Its ideal form, as Knight suggested, implied fundamental community improvements previously denied Blacks. In practice, however, race-based planning proved to be an ineffectual strategy for Black community improvement, although it did help create the segregated city.

The 1920s brought continued efforts to fashion a legally defensible racial zoning system in tandem with comprehensive city planning. Also, race-based planning spread into new places. Birmingham, Alabama, was one of the new converts to racial zoning as well as the broader version of race-based planning. Although the city lacked an official planning body, it hired Warren Manning, a Boston landscape architect, as its planning consultant and released the "City Plan of Birmingham" in 1919. The Manning plan offered a series of general recommendations about land use, transportation, and civic improvements for a city that during the previous decade increased its land area sevenfold and its population by 150,000 persons.[37]

Although slow to embrace the full range of planning proposals, especially the expensive items such as a new civic center, parks, and new roadways, city commissioners quickly recognized the need to control land uses in the midst of hectic development. In 1925, the city enacted a modified racial zoning ordinance "to protect the property holders against manufacturing plants and corner grocery stores which tend to spring up promiscuously about the city and to restrict the negroes to certain districts."[38] As late as 1926, Birmingham's zoning system provided for the rigid racial separation of residential areas, although it permitted property ownership by one race in districts allocated to members of the other race. The city commission used its power to issue or revoke building permits to prevent "construction of Negro housing contiguous to White neighborhoods." Racial zones dictated Birmingham's residential development patterns from 1926 to 1949.[39]

Robert Whitten's Atlanta Zoning Plan of 1922 was a prominent post-*Buchanan* attempt to link legalized residential segregation to comprehensive planning. Actually, what Whitten proposed differed little from the City's original "unconstitutional" racial zoning scheme, except that it employed the nomenclature of conventional zoning along with racial designations such as: R1—White district; R2—colored district; and R3—undetermined. Whitten defended racial zoning on the grounds that the Atlanta plan allowed "adequate areas for the growth of the colored population," that residential separation would instill in Blacks "a more intelligent and responsible citizenship," and that racially homogeneous neighborhoods promoted social stability. Even in its new guise, Atlanta's racial zoning plan failed to survive its initial court challenge.[40]

This renewed attempt to institute racial zoning took place within the context of a major metropolitan planning initiative, under the guidance of planning consultant Warren Manning, to make Atlanta "a beautiful, orderly place, the wonder city of the southeast."[41] Even though the explicit racial designations in the city's zoning ordinance had to be excised, Atlanta still pursued the "controlled segregation" objective of race-based planning over the ensuing decades. According to the 1922 plan, Atlanta's Black residential expansion was to be confined to the west and southwest sections of the city. That was, in fact, exactly the direction of Black residential expansion from the 1920s onward, even though the traditional heart of the Black community was in east Atlanta.[42]

Another newcomer to the racial zoning movement in the 1920s was New Orleans. Although the Crescent City discussed the implementation of racial zoning prior to *Buchanan*, not until the city secured zoning authority in 1921 did it attempt to frame an ordinance "to evade the ruling of the Supreme Court."[43] In 1923, New Orleans created an official city planning commission—being the first Southern city to do so—and quickly drafted a preliminary zoning ordinance. Two years later, the city created another advisory group, the Vieux Carre Commission, to suggest to city council ways "to protect the old colonial city from 'the encroachment of modern business.' "[44] In 1927, the city hired Harland Bartholomew to begin work on a master plan.[45]

Entwined within this sweeping set of planning initiatives was a racial zoning scheme. The New Orleans ordinance stipulated that Blacks could not occupy a house in a White block or a White person in a Black block unless the prospective occupant obtained written permission of a majority of the residents already in the block. Although sustained by a lower court, the Louisiana Supreme Court reversed the decision. The New

Orleans race-based occupancy-by-permission-slip zoning arrangement proceeded to the nation's highest court for review. It was debated on March 8, 1927, virtually in the shadow of the landmark *Euclid* decision that sanctioned zoning. A week later, the court rejected the New Orleans ordinance, citing *Buchanan v. Warley* as the guiding precedent in its decision. While the decision was not unexpected, it is interesting to note how New Orleans attempted to frame its defense in terms of planning to achieve social rearrangement, not just property protection. The New Orleans city attorney contended that racial zoning was not merely an exercise of the authority recently upheld by the Court in the *Euclid* decision but also a corollary to an earlier court decision in *Plessy v. Ferguson* (1896). New Orleans argued that zoning and comprehensive planning should join the host of legal Jim Crow strategies being employed to transform the racially integrated Southern city into a bifurcated racial world.[46]

In Charleston, South Carolina, another city that hopped onto the planning bandwagon in the 1920s, racial zoning took a backseat to historic zoning but it was, nonetheless, integral to the city's comprehensive plan. At the urging of the Society for the Preservation of Old Dwellings in Charleston, the city hired a planning consultant, Morris Knowles of Pittsburgh, to prepare a zoning ordinance "sensitive to the unique heritage of Charleston." In conjunction with the zoning ordinance, which was the first in the nation to contain explicit protection for a designated historic district, Knowles prepared a general plan that included recommendations for street widenings and new thoroughfares, new schools and parks, and the creation of different land use districts. The plan also delineated separate residential districts for Blacks and Whites, although explicit racial labels were left out of the official zoning nomenclature.[47]

It is significant not only that Charleston still experimented with racial zoning as late as 1931 but also that it was one of the first cities to link racial exclusion to neighborhood preservation. According to the Knowles general city plan, the area embraced by the newly created Old and Historic District, which in 1931 still contained several thousand Black residents, was to become White. The testimony of local preservationists indicates that displacement of Blacks from the historic area was one of the implicit goals of the plan and a desired outcome of neighborhood revitalization.[48]

John Nolen's 1928 comprehensive city plan for Roanoke, Virginia, provides another example of how Southern cities, with the assistance of their planning consultants, dealt with zoning as a social control device in the aftermath of the U.S. Supreme Court's strictures against racial zoning. Nolen's 1928 plan constituted an expanded version of his 1907 "beautification" plan for Roanoke. Nolen acknowledged in his update that "Roanoke has shown inclination to include negroes of the city when making improvements for the betterment of the city as a whole."[49] Unlike Whitten in Atlanta, however, Nolen did not recommend explicit racial zoning, not simply because it would fail legally but also because its intent seemed to be a *fait accompli* by the late 1920s. Blacks in Roanoke were already segregated. As indicated in his Map of Existing Conditions, "negro residences" were concentrated south of Washington Park, with only a small pocket of Black households congregated adjacent to the Norfolk and Western railroad tracks to the west of the downtown.

Nolen dealt with the rationale behind residential segregation in a separate, one-page (two-paragraph) section of the plan titled "Areas for Colored Population." Here he noted that "zoning will protect their homes from the encroachment of business and industry in the same manner as in all other sections of the city."[50] As to what land uses would be allowed in existing Black neighborhoods, the 1928 plan remained silent. The "Existing Conditions" map treated Black neighborhoods as "special" areas without reference to the sort of land use classification scheme employed in White areas. Rather than establishing explicit racial residential zones, Nolen noted only that "general expansion (of Black residences) coordinating with that of the whole city will be an important part of the city planning program."[51]

One further example of the initial post-*Buchanan* approach to racial zoning can be seen in another Nolen project, the planning of the resort city of Venice, Florida, in the mid-1920s. While the impetus for Venice was the Florida coast land boom of the 1920s, and the desire to market property to the state's affluent new migrants, as a Southern new town, Nolen had to make a place for Black residents. As Nolen noted,

> In all Southern developments adequate provision for the negro working population is of great importance. . . . The only satisfactory answer is the setting aside of a tract large enough (and yet not too large), and planning it completely for negro village life.[52]

Nolen first outlined his rationale for a separate "village" in his plan for Kingsport, Tennessee, a decade earlier. The Kingsport plan included

> a negro village of a high order with their own schools, churches, lodges, etc., providing the same grade of housing and general development as is furnished the White population of the same economic condition.[53]

Nolen regarded Kingsport's Armstrong Village as an alternative, as he put it, "to the squalid 'Nigger-districts' so common in Southern communities."[54]

Nolen resurrected the village scheme for his work in Venice, proposing a community of substantial house lots (50 by 200 feet) and space for other amenities of community life, including a small park, a village square for stores and community buildings, four church sites, a community school, and a swimming lake. Although the developers of Venice had Nolen reduce the size of the house lots and increase their number, the village was never built. The fate of the Negro Village for Venice may be explained by the general condition of many plans prepared by Nolen and others involved in the Florida land boom. As John Hancock put it so aptly, "City planning in Florida typified the twenties dilemma of not wanting to be without a plan and not wanting to do anything about it once made."[55]

When it came to implementing plans for model Black communities in Southern cities, the dilemma was not quite so profound. As an alternative to merely segregating Blacks in the least attractive sections of existing towns and cities, the Negro village or new community approach seemed too grandiose and, hence, unnecessary.

BEYOND RACIAL ZONING:
PLANNING IN THE 1930s AND 1940s

By the 1930s, the racial zoning movement had run its course. This is not to suggest that the racial imperatives of zoning disappeared, however. The 1930s and 1940s constituted an important period in local planning since many cities had just recently devised plans that called for separate Black sections regulated in various ways. Federal initiatives in public housing and slum clearance provided additional resources for reconstructing the social landscape, and Southern and non-Southern cities eagerly participated in these efforts. With these new tools for social engineering, Southern cities ceased to confront head-on the legal objections outlined in the *Buchanan* decision. In Virginia, for example, the final court test of Richmond's racial zoning plan occurred in 1929 when a Black property owner brought suit after being denied access to a house he owned in a "White" neighborhood. Resting squarely on the 1917 Supreme Court decision, a federal circuit court of appeals upheld a lower court ruling that the city's ordinance was intended to restrict property use on the basis of race and declared the city zoning law unconstitutional.[56] A state court struck down a general zoning ordinance in Winston-Salem, North Carolina, which provided for racial districts. Birmingham continued illegally to enforce a racial zoning code until 1951.[57]

The substitute for racial zoning was a race-based planning process that marshaled a wide array of planning interventions in the service of creating separate communities. Street and highway planning served as a means to erect racial barriers as early as the 1920s.[58] The siting of public housing projects explicitly (and legally) for Black occupancy proved particularly effective in furthering residential segregation. Slum clearance, neighborhood planning, private deed restrictions, and racially charged real estate practices all served the cause of segregation as effectively as racial zoning. As the planning movement abandoned efforts to create a legally defensible system of racial zoning, support for race-based planning moved outside the Southern region.

All of the nation's major cities, especially those outside the South, experienced huge increases in Black population after 1940. The ensuing battle between Blacks and numerically declining Whites for space in the center city produced "powerful social consequences." While it may be too much to argue that the national urban planning movement was consumed by racial issues beginning in the 1920s, it is fair to suggest that a widely shared underlying premise of planning was the need to pursue community improvements within the context of separate racial worlds. Planners such as Nolen proposed opportunities to realize substantive community improvements under the aegis of apartheid, but rarely were these ideals realized. Indeed, as the planning movement expanded beyond racial zoning to "racially informed comprehensive planning," it became more difficult to distinguish between planning for general community improvement and planning merely to reinforce and sharpen "the color line." The widespread practice of communities seeking to exclude "undesirables" through exclusionary zoning had its greatest impact on African Americans (see Ritzdorf, Chapter 3).[59]

Both public housing and urban renewal exemplified the difficulty of positively addressing the problems of the Black community within the murky context of race-based planning. Black leaders and citizens were wary of, and in some cases openly hostile toward, low-income housing projects as early as the 1930s, even though reformers and planners maintained that an impoverished group was being offered substantially improved housing and community facilities. Of course, some opposition by African Americans stemmed from expropriation of land from some Blacks to build public housing or carry out the slum clearance. Still, the land expropriation cannot explain the depth of the hostility to an otherwise legitimate community improvement effort.[60]

The explanation may lie in the experiences of the previous two decades, in which the motives for planning, not only in the urban South but also in the inner-city communities of the North, became clearly associated with the prerogatives of race-based planning. The legacy of the racial zoning movement had an enduring influence on how Blacks perceived the methods and intentions of city planning. In urban communities throughout the South, beginning in the 1930s, Blacks quietly but decidedly launched a tradition of challenging the ideas of their supposed benefactors, the city planners. At times, the planning legacy of racial zoning may have blinded Blacks to the benefits of certain community projects and planning approaches. As African Americans emerged as a dominant social and political force in American cities in the 1960s, planners quickly discovered that they had cultivated some rather strident opponents. Among their Black clients, in particular, the social legacy of the early zoning movement lived on in the politics of urban America, and challenged planners to adopt a more inclusionary approach to urban development.

NOTES

1. Mel Scott, *American City Planning Since 1890* (Berkeley, CA: University of California Press, 1971), 75-76; Mark Weiss, *The Rise of the Community Builders: The American Real*

Estate Industry and Urban Land Development (New York: Columbia University Press, 1987), 81-85; Charles Mulford Robinson, *City Planning* (New York: O. P. Putnam's Sons, 1916), 280.

2. Charles M. Haar, "Reflections on Euclid: Social Contract and Private Purpose," in *Zoning and the American Dream: Promises Still to Keep,* ed. Charles M. Haar and Jerold S. Kayden (Washington, DC: American Planning Association Press, 1989), 337.

3. Yale Rabin, "Expulsive Zoning: The Inequitable Legacy of Euclid," in Haar and Kayden, *Zoning and the American Dream,* 105; Harvey A. Kantor, "Benjamin Marsh and the Fight over Population Congestion," *Journal of the American Institute of Planners* 40 (1974): 422-29; Haar and Kayden, *Zoning and the American Dream,* 349.

4. Rabin, "Expulsive Zoning," 106.

5. Ibid., 101.

6. Gilbert T. Stephenson, "The Segregation of the White and Negro Races in Cities," *South Atlantic Quarterly* 13 (1914): 1-18; Roger L. Rice, "Residential Segregation by Law, 1910-1917," *Journal of Southern History* 64 (1968): 179-99; Barbara J. Flint, "Zoning and Residential Segregation: A Social and Physical History, 1910-1940" (Ph.D. diss., University of Chicago, 1977).

7. There is excellent and growing literature on various efforts to control residential change in the Black community after racial zoning was declared unconstitutional in 1917. Michael J. O'Connor, "The Measurement and Significance of Racial Residential Barriers in Atlanta, 1890-1970" (Ph.D. diss., University of Georgia, 1977); Ronald H. Bayor, "Expressways, Urban Renewal and the Relocation of the Black Community in Atlanta," paper presented to the Organization of American Historians, Reno, NV, 1988; Raymond A. Mohl, "Race and Space in the Modern City: Interstate 95 and the Black Community in Miami," paper presented to the American Planning Association, Atlanta, GA, 1989; Arnold R. Hirsch, *Making the Second Ghetto: Race and Housing in Chicago, 1940-1960* (Cambridge: Cambridge University Press, 1983); John F. Bauman, *Housing, Race and Renewal: Urban Planning in Philadelphia, 1920-1974* (Philadelphia: Temple University Press, 1987).

8. Weiss, *The Rise,* 83-84.

9. Flint, "Zoning"; Allen H. Spear, *Black Chicago: The Making of a Ghetto, 1890-1920* (Chicago: University of Chicago Press, 1967); William M. Tuttle Jr., *Race Riot: Chicago in the Red Summer of 1919* (New York: Athenaeum, 1974); Hirsch, *Making the Second Ghetto.*

10. Blaine A. Brownell, "The Urban South Come of Age, 1900-1940," in *The City in Southern History,* ed. Blaine Brownell and David R. Goldfield (Port Washington, NY: Kennikat Press, 1977), 138.

11. Stanislaw J. Makielski, *Local Planning in Virginia: Development, Politics and Prospects* (Charlottesville: University of Virginia, Institute of Government, 1964), 175; City of Richmond, Ordinances and Resolutions, September 1910-August 1912, Richmond, 1913.

12. Weiss, *The Rise;* Seymour I. Toll, *Zoned American* (New York: Grossman, 1969).

13. Representative of the brief attention afforded to planning by New South historians is George B. Tindall, *The Emergence of the New South, 1913-1945* (Baton Rouge: Louisiana State University Press, 1967); Paul Gaston, *The New South Creed: A Study in Southern Mythmaking* (Baton Rouge: Louisiana State University Press, 1970). Indicative of the lack of attention to urban planning in the South in this formative period are Mel Scott, *American City Planning Since 1890* and John L. Hancock, "Planners in the Changing American City, 1900-1940," *Journal of the American Institute of Planners* 33 (1967): 293-95.

14. See Christopher Silver and John V. Moeser, *The Separate City: Black Communities in the Urban South, 1940-1968* (Lexington: University Press of Kentucky, 1995), 136-44.

15. Frank B. Williams, *The Law of City Planning and Zoning* (New York: Macmillan, 1922), 267. While many authors have acknowledged Baltimore's dubious distinction as the first city to enact a racial zoning ordinance, the only thorough assessment of that initiative is Garret Power, "Apartheid Baltimore Style: The Residential Segregation Ordinance of 1910-1913," *Maryland Law Review* 42 (1983): 296-301.

16. Power, "Apartheid," 301.

17. Charles Johnson, *Patterns of Negro Separation* (New York: Harper and Row, 1943); Rice, "Residential Segregation," 181-83.

18. Stephenson, "The Segregation," 3.

19. 110 Va. 749, 67 S.E. 376 (1910).

20. Makielski, *Local Planning*, 17; Power, "Apartheid," 304; Richmond Ordinances, 1911.

21. 117 Va. 692, 86 S.E. 139-48 (1915).

22. George D. Hott, "Constitutionality of Municipal Zoning and Segregation Ordinances," *West Virginia Law Quarterly* 33 (1927): 344-45.

23. See Christopher Silver, *Twentieth Century Richmond: Planning, Politics and Race* (Knoxville: University of Tennessee Press, 1984), 111-12.

24. Power, "Apartheid," 292-97.

25. Gustavus A. Weber, *Report on Housing and Living Conditions in the Neglected Sections of Richmond, Virginia* (Richmond: Whitte and Shepperson, 1913).

26. Joseph L. Arnold, "The Neighborhood and City Hall: The Origins of Neighborhood Associations in Baltimore, 1880-1911," *Journal of Urban History* 6 (1979): 22-23.

27. See Ray Stannard Baker, *Following the Color Line* (Garden City, NY: Doubleday, Page and Company, 1908); Nell Irvin Painter, *Standing at Armageddon: The United States, 1877-1919* (New York: W. W. Norton, 1987), 218-23.

28. Louis R. Harlan, *Booker T. Washington: The Wizard of Tuskegee, 1901-1915* (New York: Oxford University Press, 1983), 295-304; Edward Franklin Frazier, "Neighborhood Union in Atlanta," *The Southern Workman* 52 (1923): 437; E. Bernard West, "Black Atlanta—Struggle for Development, 1915-1925" (Master's thesis, Atlanta University, 1976); Michael L. Porter, "Black Atlanta: An Interdisciplinary Study of Blacks on the East Side of Atlanta, 1890-1930" (Ph.D. diss., Emory University, 1974); Dana F. White, "The Black Side of Atlanta: A Geography of Expansion and Containment, 1870-1970," *Atlanta Historical Journal* 26 (1982): 199-225.

29. *Carey v. City of Atlanta*, 143 Ga. 192, 84 S.E. 456 (1915).

30. *Harden v. City of Atlanta*, 147 Ga. 248, 93 S.E. 401 (1917).

31. Power, "Apartheid," 312-13; Rice, "Residential Segregation," 183-88.

32. Charles Knight, *Negro Housing in Certain Virginia Cities* (Richmond: William Byrd Press, 1927), 36-39.

33. Hott, "Constitutionality," 348-49.

34. *Virginia Municipal Review*, November 1925, 253.

35. Rabin, "Expulsive Zoning," 106-7; Rice, "Residential Segregation," 196; Mohl, "Race and Space," 26; White, "The Black Side"; Silver, *Twentieth Century*, 112; Morris Knowles, "General Plan of Charleston, South Carolina, 1931." Unpublished manuscript, Charleston City Archives.

36. Knight, *Negro Housing*.

37. Blaine A. Brownell, "Birmingham, Alabama: New South City in the 1920s," *Journal of Southern History* 38 (1972): 30; Brownell, "The Urban South," 1977, 136.

38. A. Texler Harrison, "Birmingham's Struggle With Commission Government," *National Municipal Review* 14 (1925): 662.

39. Birmingham, Alabama, "Zoning Ordinances of Birmingham," (Birmingham, 1926); Brownell, "Birmingham," 29; Bobby M. Wilson, "Black Housing Opportunities in Birmingham, Alabama," *Southeastern Geographer* 17 (1977): 51.

40. Bruno Lasker, "The Atlanta Zoning Plan," *Survey* 45 (1922); Robert Whitten, "The Atlanta Zoning Plan," *Survey* 45 (1922): 114-15; *Smith v. City of Atlanta*, 161 Ga. 769, 132 S.E. 66 (1926).

41. Warren Manning, "Atlanta—Tomorrow a City of a Million," *Atlanta Constitution Magazine*, March 3, 1922, 10.

42. White, "The Black Side."

43. Walter White, "The Supreme Court and the NAACP," *The Crisis*, May 1927, 82-83, 99-100.

44. Charles Hosmer, *Preservation Comes of Age: From Williamsburg to the National Trust, 1926-1949*, vol. 1 (Charlottesville: University Press of Virginia, 1981), 292-93; H. W. Gilmore, "The Old New Orleans and the New: A Case Study for Ecology," *American Sociological Review* 9 (1944): 392-94.

45. Blaine A. Brownell, "The Commercial-Civic Elite and City Planning in Atlanta, Memphis, and New Orleans," *Journal of Southern History* 41 (1975): 351-52.

46. White, "The Supreme Court," 99-100.

47. George B. Chapman, "Preservation and Planning: A Case Study of Charleston, South Carolina" (Master's thesis, University of North Carolina, 1963), 57-58; Knowles, "General Plan."

48. Interview with Mabel Pollitzer, Southern Oral History Collection, University of North Carolina.

49. John Nolen, *Comprehensive City Plan, Roanoke, Virginia, 1928* (Roanoke, VA: Stone Printing and Manufacturing Company, 1928), 65.

50. Ibid.

51. Ibid.

52. James Arthur Glass, "John Nolen and the Planning of New Towns: Three Case Studies" (Master's thesis, Cornell University, 1984), 309.

53. Margaret Ripley Wolfe, *Kingsport, Tennessee: A Planned American City* (Lexington, KY: University Press of Kentucky, 1987), 51.

54. Ibid., 51-53.

55. Glass, "John Nolen," 310-11; John L. Hancock, "John Nolen and the American City Planning Movement: A History of Cultural Change and Community Response, 1900-1940" (Ph.D. diss., University of Pennsylvania, 1964), 394.

56. *City of Richmond et al. v. Deans*, 37 F. 2d 712-13 (1930).

57. Major Gardner, "Race Segregation in Cities," *Kentucky Law Journal* 29 (1941): 213-19; Charles Connerly, "Planning Jim Crow and the Civil Rights Movement: The Rebirth and Demise of Racial Zoning in Birmingham," unpublished paper presented to the Association of Collegiate Schools of Planning, 1992.

58. See O'Connor, "The Measurement"; Bayor, "Expressways"; Mohl, "Race and Space"; Silver, *Twentieth Century*.

59. See Hirsch, *Making the Second Ghetto*; Silver, *Twentieth Century*; Mohl, "Race and Space"; Bauman, *Housing*.

60. Silver, *Twentieth Century*, 132-41, 153-54.

CHAPTER 3

LOCKED OUT OF PARADISE

Contemporary Exclusionary Zoning,
the Supreme Court, and African Americans,
1970 to the Present

MARSHA RITZDORF

By the time the U.S. Supreme Court legitimized zoning in 1926 as an appropriate use of the police powers reserved for the states under the constitution, homogeneous, exclusively "native-born," White single-family suburbs had already become firmly established as the American cultural ideal.[1] Prior to the 1920s, municipalities used both zoning and residential district laws to segregate residential areas by race and ethnicity (see Silver, Chapter 2). When the U.S. Supreme Court struck down blatantly racial zoning in the *Buchanan v. Warley* case in 1917, communities turned to less obvious but legal ways to create segregated living environments.[2]

While planners and historians consider the 1926 *Village of Euclid v. Ambler Realty Corporation* case a bellwether event in planning history, they do not often discuss the politics of racial and ethnic hatred that

43

framed its passage. The approval of zoning in 1926 occurred at a time in which nativist backlash was sweeping the United States in response to the millions of immigrants who were arriving in American cities. The migration of millions of Blacks from the South during and after World War I added to the swelling population and helped fuel negative responses by the White "native-born" population. There was an explosive growth of the Ku Klux Klan and an increase in lynchings and violent race riots.[3] It was little wonder that restrictive land covenants and zoning emerged as tools of White protectionism during this era.

Even in its earliest years, some critics expressed concern for the socioeconomic impacts that were the potential result of zoning. Others chose to view it purely as a property rights issue, although they acknowledged their concerns about the mixing of races in residential areas. Zoning was not a reform product; rather, it became a fundamental technique in maintaining the power of property.

For example, when the district court denied the right of Euclid, Ohio, to zone (the decision that was overturned in the famous Supreme Court decision), Judge Westenhaver found that "in the last analysis, the result to be accomplished is to classify the population and segregate them according to their income or situation in life."[4] He went on to explain that since the Court felt it had to strike down *Buchanan v. Worley,* he did not see how he could justify the less explicit prohibitions of Euclid, even though he supported racial segregation. "The blighting of property values and the congesting of population whenever the colored races or certain foreign races are so well known as to be within judicial cognizance."[5]

An examination of the papers of the justices and lawyers involved in the case have since revealed how strongly a desire for neighborhood racial and ethnic purity fueled the approval of this tool, since the 1920s were an era when the main focus of the Court was to support less government regulation and more freedoms for the rights of big business. "The immigrant is the main fiber of zoning."[6]

Historians have already examined the myriad ways in which communities sought and applied planning to the creation of segregated residential neighborhoods throughout the period from *Buchanan* until the 1960s (see Silver, Chapter 2, this volume). Newer research shows how segregated minority communities and neighborhoods established as a result of discriminatory zoning and land use practices are often denied the basic protections of existing Euclidian zoning. Examples of this misuse of zoning are most often presented in the literature concerning environmental racism and siting decisions made in regards to hazardous wastes and landfills (see Collin and Collin, Chapter 13). In summary, little doubt exists that racial and ethnic prejudice fueled the movement for, and the rapid adoption of, zoning ordinances throughout the United States.

THE SUPREME COURT REVISITS ZONING

Although a small handful of land use related cases were decided in the immediate post-*Euclid* era, the Supreme Court of the United States remained largely silent on zoning from 1929 to 1974.[7] Since zoning is a police power reserved for the states, state courts became the arbiter of its enforcement. For nearly half a century, the development of judicial standards for zoning derived from state judicial law. The preference of local authorities to enforce a purely residential pattern with no commercial or mixed uses presented no problems to the state courts during this period. They generally applied the *Euclid* principle.

Zoning was simply not seen as a civil rights issue. Even the *Buchanan* decision was based on the sanctity of private contracts, not on the civil rights of Mr. Warley, a Black property owner. The court, in other words, did not see zoning as a violation of his rights under the Fourteenth Amendment. Indeed, "The Supreme Court specifically declined to invalidate racial zoning on equal protection grounds. Recognizing that the rationale for the ordinance was the preservation of racial purity, the court found that the case did not address rights concerning the 'amalgamation of the races.' "[8] The court's decision rested squarely on the rights of property owners to dispose of their property as they see fit.

In the forty years between *Euclid* and the Supreme Court's reentry into the zoning arena, a variety of tools were used to maintain the "color line" in American suburbs, such as restrictive covenants (see Leavitt, Chapter 10, this volume), manipulation of the real estate market (see Mohl, Chapter 4, this volume), refusal to accept federal money and build publicly funded housing, or the acceptance of money only for senior citizens housing projects. In addition, localities regularly included multiple types of exclusionary tools in local zoning ordinances. Legal scholar Norman Williams identifies six major exclusionary devices. They are exclusion of multiple family dwellings, restrictions on the number of bedrooms in multiple family dwellings, exclusion of mobile homes, minimum building size requirements, minimum lot size requirements, and minimum lot width requirements.[9]

Enforcing exclusion was not a difficult task, since the rules established by the Federal Housing Agency, the Homeowners Loan Corporation, and the banking and the real estate industries all presumed a segregated society (see Ritzdorf, Chapter 5, this volume). Even the Civil Rights Acts of 1964 and 1968 and the Fair Housing Act Amendments in 1988 have had little impact, although they forbade discrimination in publicly funded programs and established rules for the nondiscriminatory sale and rental of housing. The 1988 amendments added language prohibiting discrimination against households with children, except in bona fide senior citizen housing projects, and protecting the rights of the disabled to accessible housing. But these pieces of legislation do not require municipalities to have or accept, for example, any subsidized or private multifamily housing. Neither of the acts in the 1960s scrutinized zoning for its impacts on the provision of fair housing, and the 1988 amendments do not expressly apply to zoning. So far, the impact of these acts on zoning has been minimal, as they apply only to issues of racial exclusion (see Planning-Related Documents, Part 5.H, this volume).[10]

Despite many efforts in the 1960s and 1970s to employ advocacy or equity planning principles (see Krumholz, Chapter 7, this volume), despite neighborhood planning to help low-income African Americans to preserve and improve their neighborhoods, despite programs like Model Cities (see Thomas, Chapter 9, this volume), and despite the actions of Davidoff and others to "open up the suburbs," African Americans remained locked out of the suburbs and locked into rigidly segregated neighborhoods.[11]

Unfortunately, when the Supreme Court reentered the area of zoning law in 1974, it provided a significant proportion of communities with more power to entrench themselves as "White only" spaces. The court made it clear that the law would protect exclusionary White suburbs from residential integration. "Discriminatory zoning practices have created and perpetrated separate residential communities for African Americans. Although segregation in employment, public accommodations, education, and other aspects of American life have lessened somewhat, residential separation in 1990 has remained essentially at the same levels as in the 1960s."[12] Even those middle-class African Americans who now live in suburbs are, for the most part, segregated.[13]

The zoning case that the Supreme Court accepted in 1974 determined the legitimacy of a very restrictive definition of family as included in a local zoning ordinance. Over the strenuous dissents of Thurgood Marshall and William Brennan, the Court supported restrictive family definitions as a legitimate use of the zoning power in *Village of Belle Terre v. Boraas* (see planning-related documents, Part 5.E, this volume). The Court went on to establish (in a set of cases discussed later in this chapter) additional legitimacy for zoning tools that are invisibly discriminatory, and it provided communities with the power to zone in racially discriminatory ways. As a result of these cases, after the end of the 1970s it became nearly impossible to use the Constitution as a wedge to end residential segregation.[14]

The following sections explore three key concepts: (1) who is a family member and concurrently who has the right to live with each other, (2) who has the right to bring a lawsuit against a community that is pursuing a discriminatory land use policy, and (3) how it is possible to prove that the intent of a municipal planning or zoning policy is racially discriminatory, if the only evidence you have is the discriminatory impact of the policy. The conclusion explores an array of other ways in which zoning can provide barriers to integrated communities. It is the intent of this chapter to provide the reader with an understanding of the social meaning of these issues, not to provide a complex legal analysis. As Perin says in commenting on the value-laden constructs of land use planning and law, "these beliefs and categories reveal much about our social structure and our moral world."[15]

WHAT IS A FAMILY?

The first two cases of importance relate to the way in which *family* is defined in zoning ordinances. In the 1960s and 1970s, family form became an issue primarily because of the increase in "nontraditional" living arrangements among young White Americans and because of the deinstitutionalization of the mentally ill.[16] Moreover, it quickly became evident that family definitions would have significant impacts on families of color as well. As in ordinances all over the United States, the town of Belle Terre placed numerical limits on the number of unrelated individuals living in a household, while allowing unlimited numbers of those related by blood, kinship or marriage. In Belle Terre, only two unrelated people were allowed to live together. The owners of a house rented to six students who attended a nearby college, thereby challenging the restriction. The District Court upheld the restriction. The Court of Appeals reversed, and the matter went to the Supreme Court.[17]

After neatly summing up all the constitutional grounds on which the *Belle Terre* case could have been adjudicated, the Court decided not to deal with any of them. In finding that the ordinance did not violate the appellants' fundamental rights of travel, association, or privacy, the Court required only that the ordinance bear a "rational relationship to a permissible state objective."[18] Justice Douglas found a permissible state objective in the village's desire to preserve traditional family and environmental values.

> A quiet place where yards are wide, people few and motor vehicles restricted are legitimate in a project addressed to family needs. The police power is not confined to elimination of filth, stench and unhealthy places. It is ample to lay out zones where family values, youth values and the blessings of quiet seclusion and clean air make the area a sanctuary for people.[19]

Belle Terre rejected the argument that unrelated individuals have a constitutionally protected right to choose their living companions and place of residence. Single-family zoning now had, as a legitimate objective, the protection and encouragement of the institution of the nuclear family. Justice Marshall vigorously objected, writing in dissent about his concerns that zoning was being used to intrude on people's First Amendment rights of association:

> Zoning officials properly concern themselves with the uses of land with, for example, the number and kind of dwellings to be constructed in a certain neighborhood or the number of persons who can reside in these dwellings. But zoning authorities cannot validly consider who those persons are, what they believe, or how they choose to live, whether they are Negro or White, Catholic or Jew, Republican or Democrat, married or unmarried.[20]

Within a year of the *Belle Terre* decision, the Court again had an opportunity to examine a family definition in a case with clear racial overtones. *Moore v. City of East Cleveland* presented the opportunity to speak to the Court on the rights of extended (but related) families.[21]

The zoning ordinance in the town of East Cleveland (a majority African American community) contained a severely limited definition of family.[22] The significance of the ordinance was the fact that it placed limitations on the rights of related family members to live together. The East Cleveland Housing Code provided that no more than one adult child and his or her spouse and children could be considered as members of the family and legitimate members of the household. Striking down the East Cleveland ordinance in a five to four decision, the Supreme Court extended the same constitutional protection to the extended family that it denied to unrelated individuals in the *Belle Terre* case. The *Moore* case is important because it does establish a constitutional right of choice of family living arrangements that local zoning authorities cannot treat lightly.

Writing an opinion joined by Justice Marshall, Justice Brennan explicitly connects the decision to the needs of African American families. Seeing the mission of the Court as one of protecting minority rights and opinions against the "tyranny of the majority," he asserted that although the nuclear family was the model found in the overwhelming majority of White households, the Constitution does not allow the government to impose that model on those who do not wish to, or cannot, live in such an arrangement.[23] He recalled the enduring tradition in immigrant families of sharing resources to survive and, as if he were writing today, pointed to the necessity for such arrangements in many contemporary families.

> The extended family form is especially familiar among Black families. We may suppose this reflects the truism that Black citizens like generations of White immigrants before them have been victims of economic and other disadvantages that would worsen if they were compelled to abandon extended for nuclear living patterns . . . the prominence of other than nuclear families among ethnic and racial minority groups including our Black citizens, surely demonstrates that the "extended family" remains a vital tenant of our society."[24]

Most family definitions simply do not contain the limitations on extended families that the East Cleveland ordinance did. Most, however, do contain limitations or complete prohibitions on the living arrangements of families composed of nonrelated individuals. In the case of African American families, this limitation can impact the creation of families based on fictive kin and the sharing of households by, for example, two unrelated single mothers and their children. In these cases, *Belle Terre* remains the controlling federal case allowing communities to go as far as they choose in forbidding the formation of alternative families.

In Missouri, for example, the State Supreme Court supported a family definition that forbids any unrelated people (including opposite-sex, nonmarried partners) from living together. In that case, *City of LaDue v. Horn*, the plaintiffs were upper-class White professionals who lived together with their children from prior marriages. The court acknowledged that in every way except for the possession of a marriage license, the couple's lifestyle was the same as a biologically related family. In addition, since neither economic or racial issues were involved, it is absolutely clear that the Court's decision was based on the members' personal moral values.[25] This decision and the reliance on *Belle Terre* have implications for many poor people, especially women and children, as they strive to create workable households and to move to neighborhoods with better educational and employment opportunities.

The ongoing entrance of new immigrant groups brings to the United States continuing racial, ethnic, cultural, and class differences in family behaviors and forms. Yet American land use patterns and zoning laws emphasize a homogenization of culture. "The tension between the ideals of family behavior in the dominant culture and the traditional patterns of the Black family and of immigrant families has been a recurring issue in American life."[26]

WARTH V. SELDIN: *WHO HAS STANDING?*

When poor and/or minority residents seek to move to suburban locations (where, more and more, the only jobs they are qualified for are locating), they find the opportunity effectively closed. They cannot find affordable housing in communities that are within reasonable commuting distance of or in the town with the available jobs. In the 1975 *Warth* case, several groups with a concern over the lack of affordable housing challenged the use of exclusionary techniques to prevent the construction of low-income housing. However, the merits of their arguments were never heard by the Court, which refused to grant them standing to bring the suit in the first place. This effectively closed the door on the right of potential residents to challenge exclusionary zoning.

Simply defined, standing is the question of whether or not a litigant has a sufficiently personal stake in the outcome of a case to warrant its consideration by the Court. The plaintiffs in *Warth* were low-income residents of communities in and around Rochester, New York, and local building and housing organizations. They sued the town of Penfield. Each group had different reasons for bringing the suit. The low-income residents asserted that their exclusion from Penfield forced them to live in less attractive environments and violated their constitutional rights to travel and to free association. The facts of the case provided overwhelming evidence that Penfield was doing everything it could to prevent the construction of multifamily housing. For example, only 0.3 percent of land available for residential construction was zoned for multifamily units. The density was so low, and other requirements so strict, that if any units were built, they would not be defined as affordable.[27]

The Supreme Court denied standing to all litigants. Justice Powell, writing for the Court, acknowledged that the allegations of the low-income plaintiffs were true and that Penfield indeed had deliberately excluded "persons of low and moderate income, many of whom are members of racial or ethnic minority groups."[28] However, the Court declined to hear the substance of the case. They claimed that since the low-income plaintiffs were not necessarily the same low-income plaintiffs who would be excluded, they could not have standing, because they could not show they had been personally injured. "The fact that these petitioners share attitudes common to persons who may have been excluded from residence in the town is an insufficient predicate for the conclusion that petitioners themselves have been excluded and that [Penfield's] assertedly illegal actions have violated their rights."[29]

Justice Brennan, writing in dissent for himself, Justice White, and Justice Marshall, wrote a blistering critique of the Court's real agenda as he saw it: a reluctance to hear on the merits a case in which they would have to have decided the constitutionality of "practices which assertively limit residence in a particular municipality to those who are white and relatively well off."[30] After examining the evidence presented to the Court regarding Penfield's zoning practices, he charged that "the purpose of this ordinance was to preclude low and moderate income people and non-whites from living in Penfield."[31] In concluding his dissent, he made a comment that was extraordinarily prescient regarding the future efforts of low-income and minority group members to challenge the constitutionality of zoning exclusion: "Understandingly, today's decision will be read as revealing hostility to breaking down even unconstitutional zoning barriers that frustrate the deep human yearning of low-income and minority groups for decent housing they can afford in decent surroundings"[32] (see Part 5.F, this volume).

METROPOLITAN HOMEBUILDERS: DISPROPORTIONATE IMPACT VERSUS INTENT TO DISCRIMINATE

During the 1960s, civil rights litigation was, as Paul Brest comments, primarily "a good guys/bad guys issue for the Supreme Court presided over by Earl Warren and there was no doubt which side the court was on."[33] Issues of racial discrimination were almost always overt and easily exposed, whether practiced by landlords, realtors, school boards, or voting registrars. By the 1970s, with a much more conservative Chief Justice (Warren Burger) presiding, two important issues related to land use began to emerge for the Court. These were less overt actions and as a result were more open to judicial interpretation.

The first set of these cases is not discussed in this chapter, although it is critical to the future of African American children. These are the cases that concern the issues surrounding the desegregation of urban/suburban school districts. Generally, the Court struck down efforts to desegregate that involved the involuntary crossing of municipal boundaries. They rejected the notion that regional solutions to discrimination in the schools were constitutionally supportable, and they essentially closed the door on urban children's access to the same quality of education as their suburban neighbors.[34]

The second set of cases addressed the legality of actions (such as that of the town of Penfield in the *Warth* case) that have a racially discriminatory effect in the absence of evidence that the rules were made with a discriminatory purpose in mind. Zoning is a perfect case of such a potential situation, and in 1977 the Supreme Court considered such a case in *Metropolitan Housing Development Corporation v. Village of Arlington Heights.*

A complex case, *Arlington Heights* has been amply reviewed in the literature on planning law.[35] In this case, the plaintiffs challenged an all-White community's refusal to rezone for racially integrated low- and moderate-income housing. The Court readily acknowledged that Arlington Heights' refusal to rezone for the proposed project had a discriminatory effect. "Since Lincoln Green would have to be racially integrated in order to qualify for federal subsidization, the Village's action in preventing the project from being built had the effect of perpetuating segregation in Arlington Heights."[36]

However, drawing on their earlier decision in an employment discrimination case (*Washington v. Davis*), they decided that the village had acted legitimately and that "there is no evidence that its refusal to rezone was the result of intentional racial discrimination."[37] They did remand the case back to the district court to consider it under the U.S. Fair Housing Act, but the parties settled out of court before the district court could reconsider the case.

Many scholars of zoning law point to the victories in various state courts to downplay the importance of these U.S. Supreme Court decisions. In New Jersey, New York, and California, decisions make it illegal for communities to place numerical limits on unrelated people in their family definitions or to treat these alternative family groups in any way that is different from the way they treat related groups for purposes of zoning[38] (see Ritzdorf, Chapter 5, this volume). In states with statewide land use planning guidelines (such as Oregon and Florida), broadly written language guarantees regional access to affordable housing, and in one state, New Jersey, in a sweeping and very important decision (*Mount Laurel*), the courts jumped into the process of eliminating local zoning that excludes affordable housing.[39]

The *Mount Laurel* case has generated much legal and planning analysis. Experts, however, do not see even this case as the basis for a major reversal of the effects of exclusionary zoning.[40] The Supreme Court's decisions are the only ones that automatically must be considered in every state. "A state high court opinion, no matter how well reasoned and persuasive, is binding only in the state in which it is delivered. U.S.

Supreme Court opinions are literally the law of the land and consequently preclude the possibility of other approaches."[41]

In plain language, this means that as we approach the twenty-first
century, African Americans' ability to challenge exclusionary zoning as
a violation of their constitutional rights is virtually nonexistent. Only if
they can make a case under the existing fair housing statutes will their
voices be heard.

> The Supreme Court has substantially limited attacks on exclusionary
> zoning based on claims of racial discrimination under the equal
> protection clause. Wholesale attacks on exclusionary zoning brought
> by non residents are foreclosed by *Warth*. *Arlington Heights* indicates
> that the Court will uphold site-specific racial discrimination claims
> only in blatantly racial cases. A municipality can apparently zone its
> entire area for single-family development and defend a refusal to
> rezone for multifamily development as consistent with its 'zoning
> factors.' "[42]

OTHER ZONING BARRIERS
IN TODAY'S COMMUNITIES

In addition to the exclusionary actions described earlier, communities
employ many other zoning tools that are often not recognized as exclusionary mechanisms. One such concept, coined "expulsive zoning" by
Yale Rabin, is the systematic use of Black neighborhoods as dumping
grounds for locally unwanted land uses.[43] These uses are most often
nonresidential in character (such as landfills), or they may be residential
uses (drug rehabilitation units) that other neighborhoods with more
political clout refuse to house. The cases that have come to court that
might fit the label of expulsive zoning have been litigated as environmental justice issues. In no case so far has a community won relief from
siting practices that, it argued, were racially discriminatory (see Collin
and Collin, Chapter 13, this volume). In several of the cases, the same
intent and impact arguments that the Court's majority used to preclude
the rezoning in *Arlington Heights* were used to deny relief. None of these
cases reached the U.S. Supreme Court.

There are also other ways in which zoning can make it difficult for
low-income families to survive in a community. Some examples include

the prohibition of, or strict limitations on, the provision of child care or elder care (especially in residential areas) and the prohibition of home work. In many cases, communities explicitly forbid the types of home work that are practiced by lower-income residents (such as beauticians and barbers), while allowing the types of home work more often practiced by the middle class (insurance or accounting, for example). The problems of undue noise, traffic, or inappropriate uses of property that are often given as reasons for these prohibitions are easily handled through nuisance and noise laws that are applied equally to all community residents regardless of race, sex, or familial relationship.[44]

Seventy years have passed since the *Euclid* decision, and exclusionary zoning is widely recognized as a major social and political issue. Yet the Congress and the courts overwhelmingly continue to view zoning as a local matter. In the most Supreme Court cases decided since 1985 in regards to zoning, the law has been used to clarify the boundaries between private property rights and acceptable land use regulation. The decisions in these cases grant privilege to the rights of the private property owner and do not indicate any interest in addressing the more global issues of zoning, especially its social and economic impact on communities at large, or the rights of those who cannot afford to own property (ownership is significantly less common for both people of color and female-headed households).

> The argument that property is a "fundamental" right and therefore requires heightened judicial protection from unjustified and frequently assumed attempts at majoritarian [in this case public regulations] invasion has been ominously revived by the *First English* and *Nollan* cases.[45]

In the meanwhile, jobs continue to move to the suburbs. The poorly housed, unemployed African American who lacks adequate, affordable housing within reasonable distance of these jobs has little cause for optimism that segregated suburban communities will change their practices any time in the near future. Ironically, changes in federal housing policy that now emphasize consumer mobility through the use of housing vouchers assume that voucher recipients can use their vouchers to improve their locational choices without taking into account the problems that zoning creates in providing residential choice.

A recent study of eighteen housing mobility pilot programs concluded that increased availability of units in more affluent neighborhoods is critical if the programs are to be successful.[46] Therefore, as long as zoning continues to be used as a subtle, potent tool of segregation, mobility programs such as this will not succeed.

Zoning spatially allocates wealth, prestige, and opportunities in American communities. It is a very effective way for communities to create legal barriers that support a hierarchy in which some human beings are privileged and others are subordinated because of their class, race, and gender characteristics.

NOTES

1. *Native-born white* is the term that was used by the White Christian middle class whose ancestors were primarily Dutch, Scotch, English, and German and had arrived in the earliest waves of immigration to describe themselves around the turn of the century.

2. *Village of Euclid v. Ambler Realty Corporation*, 272 U.S. 365 (1926); *Buchanan v. Warley*, 245 U.S. 60 (1917).

3. Nicholas Lemann, *The Promised Land: The Great Black Migration and How It Changed America* (New York: Knopf, 1991).

4. *Euclid v. Ambler*, 312-13.

5. Ibid.

6. Seymour Toll, *Zoned American* (New York: Grossman Publishers, 1969), 158.

7. Richard Babcock, *The Zoning Game* (Madison: University of Wisconsin, 1966).

8. Jon C. Dubin, "From Junkyards to Gentrification: Explicating a Right to Protective Zoning in Low Income Communities of Color," *Minnesota Law Review* 77 (1993): 746.

9. Norman Williams, *American Planning Law* (Chicago: Callaghan, 1975).

10. In 1988, Congress added discrimination against families with children and the handicapped to the types of discrimination covered by the Fair Housing Act (42 U.S.C. §§ 602(k) and 8604(f). The amendments do not expressly apply to zoning but the courts have used it to invalidate ordinances that forbid group homes for the disabled.

11. Douglas Massey and Nancy Denton, *American Apartheid: Segregation and the Making of the Underclass* (Cambridge, MA: Harvard University Press, 1993).

12. Dubin, "From Junkyards to Gentrification," 755.

13. Massey and Denton, *American Apartheid*.

14. Kenneth Perlman, "The Closing Door: The Supreme Court and Residential Segregation," *American Institute of Planners Journal* 44 (1978): 160-69; Laurence Tribe, *American Constitutional Law* (Mineola, NY: Foundation Press, 1978).

15. Constance Perin, *Everything in its Place: Social Order and Land Use in America* (Princeton, NJ: Princeton University Press, 1977), 3.

16. Marsha Ritzdorf, "Challenging the Exclusionary Impact of Family Definitions in American Municipal Zoning Ordinances," *Journal of Urban Affairs* 7 (1985): 15-25.

17. *Village of Belle Terre v. Boraas*, 416 U.S. 1 (1974), 94 S. Ct. 1536.

18. Ibid., 1540.

19. Ibid., 1541.

20. Ibid., 1543.

21. *Moore v. City of East Cleveland,* 431 U.S. 494 (1977).

22. Ibid. The complete definition is not reprinted here in the interest of space but can be found in the case.

23. Charles Harr and Jerold Kayden, *Landmark Justice: The Influence of William J. Brennan on America's Communities* (Washington, DC: Preservation Press, 1989), 50.

24. Ibid., 123.

25. *City of LaDue v. Horn,* 720 S.W. 2d 745 (Mo. App., 1986).

26. Maxine Baca Zinn, "Family, Feminism and Race in America," *Gender and Society* 4.1 (1990): 71.

27. *Warth v. Seldin,* 422 U.S. 490 (1975).

28. Ibid., 500-506.

29. Ibid.

30. Ibid., 528-29.

31. Ibid.

32. Ibid.

33. Paul Brest, "Race Discrimination," *The Burger Court,* ed. Vincent Blasi (New Haven, CT: Yale University Press, 1983), 113.

34. See, for example, *Milliken v. Bradley,* 418 U.S. 717 (1974).

35. *Metropolitan Housing Development Corporation v. Village of Arlington Heights,* 558 F. 2d 1283 (7th Cir., 1977).

36. Ibid.

37. Ibid.

38. Ritzdorf, "Challenging the Exclusionary Impact," 15-25.

39. *Southern Burlington N.A.A.C.P. v. Township of Mount Laurel,* 456 A.2d 390 (1983).

40. See, for example, Peter Abeles, "Planning and Zoning," in *Zoning and the American Dream,* ed. Charles Harr and Jerold Kayden (Chicago: Planners Press, 1989), 147-50.

41. Harr and Kayden, *Landmark Justice,* 16.

42. Daniel Mandelker, *Land Use Law,* 3rd ed. (Charlottesville, VA: Mitchie Company, 1993), 322-23.

43. Yale Rabin, "Expulsive Zoning: The Inequitable Legacy of Euclid," in Harr and Kayden, *Zoning and the American Dream,* 101-21.

44. Marsha Ritzdorf, "A Feminist Analysis of Land Use Planning and Zoning," in *Women and the Environment,* ed. Irwin Altman and Arza Churchman (New York: Plenum Press, 1994), 255-79.

45. Robert A. Williams Jr., "Euclid's Lochnerian Legacy," in Harr and Kayden, *Zoning and the American Dream,* 294.

46. Alex Polikoff, *Housing Mobility: Promise or Illusion* (Washington, DC: Urban Institute Press, 1996).

CHAPTER 4

THE SECOND GHETTO AND THE
"INFILTRATION THEORY" IN URBAN
REAL ESTATE, 1940-1960

RAYMOND A. MOHL

In the years between 1940 and 1960, powerful demographic and structural shifts began to reshape neighborhood life in urban America. During these two decades, some five million African Americans migrated from the South to the urban North and West. Large numbers of Southern Blacks also moved from the rural to the urban South. At the same time, and especially after 1945, millions of White Americans began moving to the sprawling tract houses of the new postwar suburbs. In many cities, the suburban migration was also, in part, a racial response on the part of Whites seeking to escape what was perceived as a Black "invasion" of central-city residential neighborhoods. The period between 1940 and 1960, therefore, was one of dramatic racial and spatial reorganization of the American metropolis. African Americans who moved into the cities pushed out the residential boundaries of the inner-city ghettos, moving into the neighborhoods of Whites who had departed for the suburbs. Once the racial transitions began, more Whites moved out because of the Black in-migration. The creation of the "second ghetto," as historians have labeled this process of racial/spatial transition, emerges in retrospect as one of the most significant structural changes in the mid-twentieth-century American city.[1]

The process of residential transformation was not an easy one. Indeed, the frontiers of neighborhood change were marked by bitter protest, picketing, boycotts, demonstrations, intimidation, bombings, arson, and mob violence in many postwar cities. Neighborhood improvement associations actively participated in the defense of White communities faced with African American "infiltration." As one research report from 1947 noted, "the function of Negro exclusion" had become "the controlling motive" for most neighborhood associations: "the maintenance of Caucasian-pure residence areas has come to be a dominant purpose." White communities mobilized to prevent the racial transition of urban neighborhoods.[2]

Propelling the process of second ghetto formation was the intensifying demand for new and better housing among Black Americans. The boundaries of the older, inner-city ghettos formerly had been maintained by racial zoning and restrictive covenants, as well as by the discriminatory policies of private lending firms. Federal housing agencies such as the Home Owners Loan Corporation (HOLC), the Federal Housing Administration (FHA), and the Veterans Administration (VA) supported lily-white housing practices, and public housing policies also adhered to the color line.[3] By midcentury, these discriminatory mechanisms could no longer contain the burgeoning Black populations of the American metropolis, now swelled by a decade of Southern Black migration. Court challenges supported by groups such as the National Association for the Advancement of Colored People (NAACP) resulted in the outlawing of restrictive covenants by the U.S. Supreme Court in 1948.[4] Many state and U.S. district courts in the South banned local racial zoning laws in the late 1940s.[5] Both the private sector and federal housing agencies began backing away from official policies of discrimination by the early 1950s, but not until the civil rights legislation of the mid-1960s was housing discrimination fully outlawed.[6] Nevertheless, throughout the 1940s and 1950s, although White neighborhood associations sought to maintain the color line in urban housing, African Americans who wanted better housing seized on opportunities presented by more liberal court decisions and by more open housing and lending policies. They were willing to challenge segregation and confront mob terror to achieve their objectives.[7]

There was one other mediating force facilitating, often manipulating, the racial/spatial transitions of the postwar American city. Indeed, central to the process of neighborhood change during these years was the real estate industry—the hidden hand that helped shape the postwar metropolis. Real estate professionals were fully cognizant, from an early period, of the demographic changes that were restructuring the city. Innumerable articles in the real estate trade journals not only demonstrated an awareness of racial change but provided prescriptive advice for real estate people confronted with racially changing neighborhoods. The responses were not monolithic by any means. Most in the real estate industry sought to hold off racial change and stabilize White neighborhoods; some others—blockbusters or builders of new Black housing, for instance—sought to profit from the demographic changes buffeting America; a very few hoped to facilitate residential integration and denied that racial change undermined property values.[8] With the growth of the second ghetto providing the background, this chapter will first discuss the Black migration and urban violence in the second-ghetto era and then outline the role and response of the real estate industry to urban change as reflected in the industry's professional and trade journals.

BLACK MIGRATION AND SECOND GHETTO VIOLENCE

As noted, the demography of American cities and metropolitan areas began changing dramatically during the 1940s and after. The process of change began with new migratory patterns during World War II, as White and Black war production workers moved to new "war-industry centers," especially in the Northeast, Midwest, and Pacific Coast states. In the postwar years, urban America experienced a convergence of regional and metropolitan migration trends, as African Americans continued to move out of the South to new urban destinations and as urban Whites increasingly migrated to the suburban periphery of the large metropolitan areas.[9]

The U.S. metropolitan population more than doubled between 1940 and 1960, from 63 million to 133 million. Virtually all of the largest

American metropolitan areas grew rapidly between 1940 and 1960; for instance, metropolitan Detroit increased in population by 64 percent, Philadelphia by 50 percent, Chicago by 38 percent, Los Angeles by 132 percent, and San Diego by 303 percent. However, almost all of that metropolitan growth was recorded in the suburbs. Many of the largest central cities (except those in the Sunbelt) had already stopped growing during these years, and some even began losing population. The trend was very clear—in mid-twentieth-century metropolitan America, population growth was a phenomenon that occurred mostly on the fringes and in the suburbs.[10]

Striking increases in the central-city Black population matched the White suburban surge. Virtually every big city in the United States recorded sharp gains in the Black population between 1940 and 1960. The big East Coast cities—New York and Philadelphia—and almost all of the industrial cities of the Midwestern heartland showed amazing Black population gains ranging from 100 percent in Philadelphia to 301 percent in Buffalo and 607 percent in Milwaukee. Even more startling was the West Coast experience; Los Angeles with a 425 percent increase was actually slow growth compared to the rate of increase in Seattle (592 percent), San Diego (721 percent), Oakland (882 percent), and San Francisco (1,425 percent). If the White population of the central cities was emptying out to the expanding suburban fringe, the cities themselves served as a massive human magnet attracting the African Americans migrating out of the South. Consequently, the proportion of Blacks in the central-city population was rising rapidly just about everywhere during these midcentury years. This experience of migration and metropolitan change established the basis for the later emergence of many majority Black central cities—the pattern most common at the end of the twentieth century.[11]

The conjunction of the African American migration to the city and the White migration to the metropolitan suburbs brought powerful social consequences. Despite large increases in Black population during the 1940s, the color line in housing continued to prevail. Severe housing congestion marked the confined Black residential areas in just about every big city and many smaller ones. Typically, as families doubled up in apartments, and as single-family residences were subdivided into multifamily units, overcrowded conditions worsened. These central-city housing pressures eventually contributed to the opening up of new areas for Black residence. The racial succession process varied in different cities, with some city governments providing new public housing for African Americans, others (especially in the South) zoning undeveloped spaces for new Black residence. Race advancement organizations such as the NAACP and the National Urban League pressured government agencies, and legal challenges produced some positive results. In most cities, some elements of the real estate industry facilitated the process of racial change through "blockbusting." But whatever the method, African Americans were eager for more and better housing, and they generally pushed the limits to achieve it. Thus, racial/spatial transitions took place in virtually every big city, as African American housing pioneers blazed new trails into White residential neighborhoods.

However, breaking the color line on housing was accompanied by great personal risk for African Americans. During the earlier "great migration" from the South to the urban North—the migration of 1915 to 1930—competition for housing and consequent racial conflicts touched off violent race riots in East St. Louis and Houston in 1917, in Philadelphia in 1918, in Chicago and Washington, D.C., in 1919, and in Tulsa in 1921. In Miami, Detroit, Baltimore, Kansas City, Memphis, and dozens of other cities, White mobs used intimidation, terror, and violence to maintain the color line in housing. The pattern of intimidation and violence surged once again in the midcentury decades, as the urban Black population began rising rapidly once again, as housing pressures intensified, and as new second ghettos sprouted all over urban America.[12]

The racial violence of the 1960s—the urban riots and ghetto insurrections of the Great Society era—is well known and much studied. Surprisingly little is known about the urban racial violence of the period between 1945 and 1960, when the victims were mostly Black people seeking better housing and the violent perpetrators were mostly White people trying to prevent African Americans from moving into their neighborhoods. Yet these racial incidents were ubiquitous at the time, even if they were not very well reported in the metropolitan press or by national news magazines. Indeed, as Herbert Shapiro has suggested, it appears that "all during this period news coverage of the racial violence was substantially suppressed." (The left-liberal press and the African American press did a much better job of reporting this surge of second ghetto housing violence.) Just barely hidden from general public awareness, however, an era of "chronic urban guerrilla warfare" was emerging as the racial dynamics of neighborhood change produced violent responses from Whites who feared, resented, and often resisted Black "intrusion" into their neighborhoods.[13]

The White residents of midcentury Chicago could not have missed what was happening in working-class southside and westside neighborhoods. According to Arnold R. Hirsch, whose book *Making the Second Ghetto* (1983) provides a case study of the racial/spatial transitions in midcentury Chicago, more than 350 incidents of racial violence related to housing were reported to the Chicago Commission on Human Relations between 1945 and 1950; Chicago's racial troubles became even more intense in the 1950s. In Detroit, according to Thomas J. Sugrue, who studied the process of racial/spatial change in the auto city, "more than two hundred violent incidents occurred in racially transitional neighborhoods between 1945 and 1965, including the gathering of angry crowds, rock throwing, cross burning, arson, and other attacks on property." In Philadelphia, according to John F. Bauman, racial incidents were occurring with regularity by the 1950s: "Some 213 racial incidents during the first six months of 1955 alone reinforced an antiblack climate that cemented the boundaries of the city's North Philadelphia ghetto." In Miami, Florida, dozens of cross burnings, house burnings, bombings, and other forms of intimidation occurred between 1945 and 1951, as Blacks pushed out the boundaries of the confined inner-city ghetto. In the early 1950s, mob violence, arson, and bomb throwing occurred in Chicago, Atlanta, Dallas, Kansas City, East St. Louis, Birmingham, Louisville, Cleveland, Philadelphia, Indianapolis, Los Angeles, Tampa, and several California cities, to name just a few such cases.[14]

The violent responses to racial/spatial neighborhood transitions knew few bounds. They occurred in Northern and Southern cities, in the East and the West, in big cities and small cities, even in smaller suburban towns. Most of these incidents, which must have totaled many thousands nationwide, followed a similar pattern, usually touched off when a Black family bought or rented in a White district. During these midcentury years, as the Black migration intensified and as city neighborhoods began to experience racial succession, White people everywhere seemed easily riled up over the thought (or the reality) of African Americans moving into their neighborhoods. Carefully watching these events, and often heavily involved in them, local real estate people had a major interest both in the defense of White neighborhoods and in the making of the second ghetto.

THE REAL ESTATE INDUSTRY
AND THE "INFILTRATION THEORY"

The real estate industry had a major interest in the residential changes that accompanied the Black migration to the postwar American city. Many local real estate sales people and brokers confined their activities to specific urban neighborhoods or districts. Economic or social forces that affected such districts had consequences for their business and income. In many cities, real estate men served as officers and leaders of neighborhood improvement associations; in other cities they worked in concert with the associations in promoting restrictive covenants and maintaining White neighborhoods.[15] Similarly, local banks, savings and loans, and mortgage firms had big investments in urban neighborhoods and kept a close watch on residential changes. Appraisal firms had to weigh the impact of racial transitions in revising and establishing property values. Builders and developers had to keep abreast of population trends and market forces to make intelligent business investment decisions. Apartment owners, landlords, and property management firms had to consider how residential transitions would affect their taxes, property values, and pricing structures.

The dominant belief in the nation's real estate industry held, as one appraisal expert put it in 1948, that "neighborhoods change, but never for the better." Real estate analysts uniformly ticked off a laundry list of explanations for the decline of residential neighborhoods. Ubiquitously present and high up on that list was the euphemistic notation: "infiltrations of unharmonious racial groups," real estate code for the movement of African Americans into new residential areas.[16] As far back as the 1920s, the National Association of Real Estate Boards (NAREB) drafted a code of ethics that made the sale of homes in White neighborhoods to African Americans a breach of professional standards. Most regional, state, and local real estate organizations replicated and adhered to the NAREB professional code. The National Association of Home Builders pursued similarly discriminatory practices, advocating home building for Blacks only in segregated neighborhoods. The NAREB deleted the racial provision of its ethics code in 1950, but the policy continued to be observed informally throughout the real estate industry. Most real estate people took it for granted that the movement of Blacks into White neighborhoods would undermine property values and destroy the community.[17]

These discriminatory perceptions were deeply embedded in the process of training and educating real estate professionals. The NAREB standards regarding race were written into virtually all of the real estate textbooks, appraisal manuals, and training materials at the time. As early as 1932, for instance, in his real estate text, *The Valuation of Real Estate*, Frederick M. Babcock of the University of Michigan devoted a full chapter to "The Influence of Social and Racial Factors on Value." Babcock advanced the concept that all residential neighborhoods gradually declined but that the downward trend in value could be speeded up by an "infiltration process" that "carries all residential communities not capable of other use downward in quality and value."[18]

In their book *Principles of Urban Real Estate* (1948), Arthur M. Weimer and Homer Hoyt also embraced the infiltration theory. Neighborhood stability, Weimer and Hoyt contended, could be maintained only "if the people living in an area are not threatened by the infiltration of people of another racial or national type." The migration of such groups into a neighborhood, they wrote, "frequently stimulates the out-migration of previous residents in the area." It is curious that those advancing the "infiltration" argument rarely mentioned Negroes or Blacks specifically, but given the real estate jargon of the time, everyone in the business knew what they meant.[19]

Even as official discrimination became less acceptable in the 1950s, real estate analysis of neighborhood change remained much the same. Henry E. Hoagland's 1955 text, *Real Estate Principles*, gave prominence to the infiltration theory. When "two or more incompatible groups" occupied any neighborhood, Hoagland wrote, "the tendency is for the group having the least regard for the maintenance of real estate standards to drive out the other group. . . . The infiltration of additional representatives of the dominant group operates to put a blight on the neighborhood."[20] Even as late as 1984, one real estate appraisal text was still discussing the "introduction of contentious groups" as a factor threatening property values and promoting neighborhood decline.[21] Despite the changing euphemisms over thirty years (from "inharmonious" to "incompatible" to "contentious"), real estate people had a pretty good sense of how the cities were changing and what particular responses they should pursue.

The real estate industry did not stand alone in the effort to maintain segregated housing patterns. Equally significant, federal housing agencies beginning in the New Deal era—the Home Owners Loan Corporation (HOLC), the Federal Housing Administration (FHA), the Veterans Administration (VA), the U.S. Housing Authority, the Public Housing Administration, and the Housing and Home Finance Agency (HHFA), among others, all accepted the basic premise of the "infiltration" theory. HOLC, for instance, established an appraisal system (eventually used by other federal agencies and by local bankers and mortgage firms), that initiated the pernicious and discriminatory policy of "redlining."[22] The FHA's *Underwriting Manual*, first published in 1938 and reissued in 1947, opposed neighborhood "invasion by incompatible racial and social groups" and advocated residential segregation as a means of maintaining community stability (see Part 5.B, this volume).[23] The FHA, according to housing expert Charles Abrams, "set itself up as the protector of the all-white neighborhood" and "became the vanguard of white supremacy and racial purity—in the North as well as the South."[24] Federal public housing projects were segregated from the beginning almost everywhere, a policy that was maintained by local housing authorities that placed many public housing projects in segregated neighborhoods, or at least in the path of the expanding Black ghetto. Similarly, government urban renewal and highway projects often seemed to pursue a vigorous policy of Black removal while spending less effort on the relocation of families whose housing had been taken for redevelopment purposes.[25] As Arnold Hirsch has noted, "the most distinguishing feature of post-World War II ghetto expansion is that it was carried out with government sanction and support."[26]

Professional and trade journals in the various real estate and housing fields also adhered to the infiltration theory. These journals covering real estate, appraisal, banking, savings and loan, mortgage, property management, residential development, and building activities all kept real estate professionals aware of current trends and patterns. One such journal, the *Real Estate Analyst*, edited by Roy Wenzlick of St. Louis, was published regularly through the entire period between 1940 and 1960. Consistent readers of Wenzlick's journal got very good reports on the demographic, economic, and structural changes that were reshaping urban America during the midcentury decades. As early as 1942, Wenzlick reported on the decentralization of the urban population, with numerous follow-up reports on urban population shifts, including the Black migration to the central cities. Detailed population tables periodically provided the changing statistics for cities and standard metropolitan areas. Over many years, *The Real Estate Analyst* hammered home the basic message of the "infiltration" theory, noting in one 1949 article on St. Louis, for example, that for many White neighborhoods "it would be only a matter of time before . . . Negro families encroached upon the districts in question." It was Wenzlick's expert advice that real estate appraisers should always carefully weigh the impact of "racial groups infiltrating a district." For Wenzlick, the problem was clear, and the solution clearer still—maintain the color line and hold back the encroaching second ghetto.[27]

The savings and loan industry had a big interest in neighborhood real estate, since most savings and loans were small institutions heavily tied to local housing markets. The major publication of the U.S. Savings and Loan League, *Savings and Loan Annals*, provided an annual snapshot of industry issues and concerns. Through the 1940s, the savings and loan people were mostly interested in the economic impact of the war on building materials, construction, and the new housing market. They were also extremely hostile to public housing and preferred programs, such as those later enacted in the Housing Act of 1949, that made it possible for private developers to assemble large parcels of inner-city land for private redevelopment. They rarely discussed Black housing and when they did so it was only in the context of the infiltration theory. Since most savings and loans generally practiced redlining and resisted the approval of mortgages in "second-grade districts," not many African Americans could get housing loans or even buy new houses.

However, in 1948, when a Black banker, Robert R. Taylor of the Illinois Federal Savings and Loan Association, spoke at the annual convention of the U.S. Savings and Loan League, he presented an alternative view. Taylor's savings and loan, he reported, had lent millions of dollars to African American homeowners in Chicago and from 1934 to 1948 had not foreclosed on a single property. Taylor's message apparently was ignored, and the issue of loans for Black home buyers never officially came up at conventions of the U.S. Savings and Loan League in the 1950s.

Rather, the savings and loaners seemed more comfortable with the position of one of their convention speakers in 1950, who argued that Americans should "forget about this civil rights program" and let the South "settle it" themselves. No one ever discussed the rage of mob violence that was unsettling urban neighborhoods in the North, the Midwest, and the West.[28]

Another industry trade journal, *The Mortgage Banker*, offers additional insight into the thinking and inner workings of the real estate industry. Published by the Mortgage Bankers Association of America (MBA) beginning in 1939, this trade journal kept its readers informed of current issues in the mortgage banking field. Many articles over the years reported on the decentralization of urban population and economic activities and the consequent expansion of mortgage opportunities in the new suburbs. There was a considerable amount of discussion of urban slums and "blighted" housing throughout the 1940s and early 1950s. The infiltration theory made its appearance here as well, beginning with a 1940 article that replicated the HOLC four-part grading system of residential neighborhoods. Under this appraisal system, as *The Mortgage Banker* reported it, neighborhoods with "infiltration of inharmonious families" made poor investment risks.[29]

But even *The Mortgage Banker* occasionally presented an alternative view on housing issues. In a 1949 article titled "The Mortgage Market Lenders Forgot," FHA Commissioner Franklin D. Richards contended that the Black housing market represented "a large untapped market" awaiting the courageous mortgage lender and builder. Richards reported favorably on numerous Black housing developments (apparently all segregated), noted that Black housing demand was high, that Black homeowners made good credit risks, and that the FHA ultimately took on the risk through the insured mortgage system. Richards also assured the mortgage banking industry that the FHA was willing to work with local planners and civic groups to find "acceptable sites" for Black housing developments. The FHA's Racial Relations Service, Richards said, was ready to advise communities embarking on such new housing projects. Despite this attention to the problem of Black housing in 1949, neither the FHA nor the MBA was willing to advocate integrated neighborhoods. The new all-Black housing developments promoted by the FHA at this time generally became part of the expanding second ghetto areas.[30]

In 1954, however, another FHA official, George W. Snowden, addressed a meeting of the MBA and offered an alternative viewpoint. Snowden reviewed the excellent credit record of African American home buyers, discounted the widely presumed negative effect of non-White occupancy on property values, and advocated a "uniform, single-standard, lending policy" that dealt equally with Blacks and Whites. That this sort of message was quite unusual in 1954 is suggested by the editorial comment in *The Mortgage Banker* that Snowden's address was "one of the most remarkable statements ever heard from an MBA rostrum."[31] Generally speaking, the Black housing market held little interest for the mortgage banking industry during the years of second ghetto expansion. Like most real estate people, the bankers were fully aware of the demographic changes taking place in the cities, but close adherence to the infiltration theory dictated their response.

Real estate appraisers had, perhaps, the best vantage point for observing and evaluating the impact of second ghetto migration on urban housing. One of the major journals in the appraisal field, *The Review of the Society of Residential Appraisers,* closely covered the big changes that were reshaping the American city: suburbanization, downtown redevelopment, wartime and postwar migration, highway building, public housing, new subdivision trends, new building techniques such as prefabricated housing, and the like. The appraisers were strongly supportive of restrictive covenants as a means of protecting neighborhoods and preserving property value. Many articles in the 1940s dealt with "blighted areas." "Blight" became another one of those housing euphemisms, and when it was applied, everyone in the appraisal field knew exactly what it meant.[32]

As early as 1940, the infiltration theory made its appearance in *The Review,* as the journal published an account of the HOLC neighborhood rating system and suggested that these standards "might profitably be adopted by other lending institutions." The now ubiquitous phrase, "infiltration of inharmonious families," was used to characterize "undesirable" neighborhoods, which of course offered poor mortgage credit risk. Since the appraisers were centrally involved in establishing property value, they played a major role in the redlining of racially changing neighborhoods.[33]

In later years, other articles in *The Review* discussed what they perceived as the unhappy consequences of "inharmonious racial encroachment." A few appraisers advocated new public and private housing specifically for African Americans—segregated, of course—as a means of taking the housing pressure off declining White neighborhoods endangered by "colored infiltration." In 1948, a California appraiser, George L. Schmutz, summed up what most appraisers believed: "Infiltrations of inharmonious racial groups, or improper uses of property are bound to have an adverse effect upon the value of all property in the neighborhood. The capable residential appraiser knows these things."[34]

A second professional journal in the appraisal field, *The Appraisal Journal*, published by the American Institute of Real Estate Appraisers, offered readers many similar viewpoints. As one writer noted in 1944, "infiltration of incompatible races has always been a red flag to the appraiser. It is of special importance now and will be more important after the war. No one thing can so quickly depress values as the beginning of a race movement." A similar case for the homogeneous neighborhood was made by Arthur A. May, who wrote graphically and less euphemistically in 1951 that once "the infiltration of the antipathetic racial group begins to gnaw at the edge, it will not be long until direct access will be had to the very core of the neighborhood itself."[35]

The infiltration theory dominated discussion in the pages of the appraisal trade journals, but occasionally a lonely alternative viewpoint would find its way into print. For instance, a direct challenge to prevailing attitudes was presented in *The Appraisal Journal* in 1951 in a hard-hitting article by Charles Abrams, a New York state housing official and a strong advocate of interracialism. Abrams fully rejected the infiltration theory and argued the case for integrated neighborhoods. He chastised appraisers for imposing their own personal judgments on appraisal matters; "appraisers venture into strange sociological waters," Abrams charged. Homogeneous neighborhoods were dull and monotonous, bad for children, contrary to American tradition, and un-American; moreover, they were unrealistic, given the demographic changes that were occurring in urban America. Finally, there was plenty of evidence that property values rose as Blacks moved into transitional neighborhoods. Abrams offered numerous solutions to the problem, including more public and private housing in planned neighborhoods for low-income people of all groups, elimination of restrictive practices in housing and real estate, and cessation of housing destruction in the name of slum clearance. And appraisers no doubt were surprised by Abrams's final admonition: "stop scaring people with statements that Negroes and Whites don't mix, and that when a Negro moves into the neighborhood it will soon be all Negro, or that if a Negro lives in the neighborhood he will inevitably marry your daughter, if not debauch her first." No euphemistic talk here. It is unlikely that Abrams had much impact on the real estate appraisers. At the very time he was writing, the neighborhood race wars were heating up dramatically in Chicago, Cicero, Miami, Atlanta, Detroit, Philadelphia, Los Angeles, Birmingham, Dallas, and many other cities.[36]

The real estate trade and professional journals mentioned earlier reflected mainstream opinion among appraisers, brokers, real estate analysts, mortgage bankers, and savings and loan people. Similar positions and arguments, and equally fervent adherence to the infiltration theory, might be found among other segments of the real estate industry. Alternative views came from some in the FHA (and articles in the FHA journal, *Insured Mortgage Portfolio*), from the public housers (and their magazine, *The Journal of Housing*), and from numerous other housing reformers, civil rights activists, and interracial advocates, especially among the churches. Nevertheless, the infiltration theory dominated the thinking and the politics of the real estate industry. The racial/spatial neighborhood transitions and the mob violence taking place at the street level in midcentury urban America simply confirmed the deep beliefs and the worst fears of most real estate people. Unfortunately, the real estate industry became part of the problem, rather than part of the solution to the nation's midcentury housing difficulties.

NOTES

1. Second ghetto conceptualization was initiated in Arnold R. Hirsch, *Making the Second Ghetto: Race and Housing in Chicago, 1940-1960* (Cambridge: Cambridge University Press, 1983). On the demographic and economic shifts after World War II, see also Kenneth T. Jackson, *Crabgrass Frontier: The Suburbanization of the United States* (New York: Oxford University Press, 1985); Robert A. Beauregard, *Voices of Decline: The Postwar Fate of U.S. Cities* (Cambridge, MA: Blackwell Publishers, 1993); Raymond A. Mohl, "The Transformation of Urban America since the Second World War," in *Essays on Sunbelt Cities and Recent Urban America*, ed. Robert B. Fairbanks and Kathleen Underwood (College Station: Texas A&M University Press, 1990), 8-32.

2. Herman H. Long and Charles S. Johnson, *People vs. Property: Race Restrictive Covenants in Housing* (Nashville, TN: Fisk University Press, 1947), 40.

3. Charles Abrams, *Forbidden Neighbors: A Study of Prejudice in Housing* (New York: Harper, 1955), 205-43; Robert C. Weaver, *The Negro Ghetto* (New York: Harcourt, Brace, 1948), 211-303; Norman Williams Jr., "Discrimination and Segregation in Minority Housing," *American Journal of Economics and Sociology* 9 (October 1949), 85-101; Christopher Silver, "The Racial Origins of Zoning: Southern Cities from 1910-40," *Planning Perspectives* 6 (1991): 189-205.

4. Clement E. Vose, *Caucasians Only: The Supreme Court, the NAACP, and the Restrictive Covenant Cases* (Berkeley, CA: University of California Press, 1959).

5. Restrictive covenants were outlawed by the U.S. Supreme Court in *Shelley v. Kraemer*, 334 U.S. 1 (1948). The Florida Supreme Court outlawed racial zoning in Miami in *State of Florida v. Wright*, 25 So. 2d 860 (1946). A federal district court banned racial zoning in Birmingham in *City of Birmingham v. Monk*, 185 F. 2d 859 (5th Cir., 1950). See also Jack Greenberg, *Race Relations and American Law* (New York: Columbia University Press, 1959), 275-86.

6. Arnold R. Hirsch, "With or Without Jim Crow: Black Residential Segregation in the United States," in *Urban Policy in Twentieth-Century America*, ed. Arnold R. Hirsch and Raymond A. Mohl (New Brunswick, NJ: Rutgers University Press, 1993), 90-91; Desmond King, *Separate and Unequal: Black Americans and the US Federal Government* (Oxford: Oxford University Press, 1995), 189-202.

7. L. K. Northwood and Ernest A. T. Barth, *Urban Desegregation: Negro Pioneers and Their White Neighbors* (Seattle: University of Washington Press, 1965); John Fish, Gordon Nelson, Walter Stuhr, and Lawrence Witmer, *The Edge of the Ghetto: A Study of Church Involvement in Community Organization* (Chicago: Divinity School, University of Chicago, 1966).

8. On this subject generally, see Davis McEntire, *Residence and Race* (Berkeley, CA: University of California Press, 1960), 175-250; and Rose Helper, *Racial Policies and Practices of Real Estate Brokers* (Minneapolis, MN: University of Minnesota Press, 1969).

9. Catherine Bauer, "Cities in Flux," *American Scholar* 13 (Winter 1943-44): 70-84; Philip Funigiello, *The Challenge to Urban Liberalism: Federal-City Relations During World War II* (Knoxville, TN: University of Tennessee Press, 1978): 3-38; Reynolds Farley, "The Urbanization of Negroes in the United States," *Journal of Social History* 1 (Spring 1968): 241-58.

10. Kenneth Fox, *Metropolitan America: Urban Life and Urban Policy in the United States, 1940-1980* (Jackson, MS: University of Mississippi Press, 1986), 51; U.S. Census of Population, 1940, tables 35 and 36; U.S. Census of Population, 1960, tables 44 and 45.

11. U.S. Census of Population, 1940, tables 35 and 36; U.S. Census of Population, 1960, tables 44 and 45.

12. For the racial violence accompanying the earlier black migration, see Elliot M. Rudwick, *Race Riot at East St. Louis, July 2, 1917* (Carbondale, IL: Southern Illinois University, 1964); William M. Tuttle Jr., *Race Riot: Chicago in the Red Summer of 1919* (New York: Atheneum, 1970); David Allan Levine, *Internal Combustion: The Races in Detroit, 1915-1926* (Westport, CT: Greenwood Press, 1976); Herbert Shapiro, *White Violence and Black Response: From Reconstruction to Montgomery* (Amherst, MA: University of Massachusetts Press, 1988).

13. Shapiro, *White Violence and Black Response*, 377; Hirsch, *Making the Second Ghetto*, 40-67.

14. Hirsch, *Making the Second Ghetto*, 52; Arnold R. Hirsch, "Massive Resistance in the Urban North: Trumbull Park, Chicago, 1953-1966," *Journal of American History* 82 (September 1995): 522-50; Thomas J. Sugrue, "The Structures of Urban Poverty: The Reor-

ganization of Space and Work in Three Periods of American History," in *The "Underclass" Debate: Views from History*, ed. Michael B. Katz (Princeton: Princeton University Press, 1993), 111-12; John F. Bauman, *Public Housing, Race, and Renewal: Urban Planning in Philadelphia, 1920-1974* (Philadelphia: Temple University Press, 1987), 161; Raymond A. Mohl, "Making the Second Ghetto in Metropolitan Miami, 1940-1960," *Journal of Urban History* 21 (March 1995): 395-427.

15. Long and Johnson, *People vs. Property*, 67-69; Abrams, *Forbidden Neighbors*, 181-90.

16. George L. Schmutz, "Sidelights on Appraisal Methods," *Review of the Society of Residential Appraisers* 14 (June 1948): 18; George A. Phillips, "Racial Infiltration," ibid., 16 (February 1950): 7-9.

17. Long and Johnson, *People vs. Property*, 58; Hirsch, "With or Without Jim Crow," 75; McEntire, *Residence and Race*, 175-98; Abrams, *Forbidden Neighbors*, 150-68; National Association of Home Builders, *Home Builders Manual for Land Development* (Washington, DC: National Association of Home Builders, 1953), 17.

18. Frederick M. Babcock, *The Valuation of Real Estate* (New York: McGraw-Hill, 1932), 86-92.

19. Arthur M. Weimer and Homer Hoyt, *Principles of Urban Real Estate* (New York: Ronald Press, 1948), 123, 129.

20. Henry E. Hoagland, *Real Estate Principles*, 3rd. ed. (New York: McGraw-Hill, 1955), 64-65, 236.

21. Byrl N. Boyce and William N. Kinnard Jr., *Appraising Real Property* (Lexington, MA: D. C. Heath, 1984), 125.

22. On the HOLC appraisal system in practice, see Kenneth T. Jackson, "Race, Ethnicity, and Real Estate Appraisal: The Home Owners Loan Corporation and the Federal Housing Administration," *Journal of Urban History* 6 (August 1980): 419-52; Raymond A. Mohl, "Trouble in Paradise: Race and Housing in Miami During the New Deal Era," *Prologue: Journal of the National Archives* 19 (Spring 1987): 7-21.

23. Federal Housing Administration, *Underwriting Manual* (Washington, DC: U.S. Government Printing Office, 1938); Robert E. Forman, *Black Ghettos, White Ghettos, and Slums* (Englewood Cliffs, NJ: Prentice Hall, 1971), 69-72; Joe R. Feagin and Clairece Booher Feagin, *Discrimination American Style: Institutional Racism and Sexism* (Englewood Cliffs, NJ: Prentice Hall, 1978), 105-7.

24. Abrams, *Forbidden Neighbors*, 229-30.

25. McEntire, *Residence and Race*, 291-346; Long and Johnson, *People vs. Property*, 69-72; and for a case study of Black removal, Raymond A. Mohl, "Race and Space in the Modern City: Interstate 95 and the Black Community in Miami," in Hirsch and Mohl, *Urban Policy in Twentieth-Century America*, 100-158.

26. Hirsch, *Making the Second Ghetto*, 9.

27. Roy Wenzlick, "Land Appraising," *Real Estate Analyst* 18 (June 10, 1949): 233-36; Roy Wenzlick, "Migratory Population Changes by Race," ibid., 23 (June 30, 1954): 260-61. On Wenzlick, see also Louise Cooper, "Real Estate's Prophet," *Freehold: The Magazine of Real Estate* 9 (May 1942): 39-44.

28. Robert R. Taylor, "Financing Opportunities Among Minority Groups," *Savings and Loan Annals, 1948* (Chicago: U.S. Savings and Loan League, 1949), 144-46; Raymond A. Moley, "The Businessman's Stake in Politics," *Savings and Loan Annals, 1950* (Chicago: U.S. Savings and Loan League, 1951), 39.

29. "Factors in the Rating of Neighborhoods," *The Mortgage Banker* 2 (October 1, 1940): 6-7; Wallace Moir, "Slums: Are We Creating Them Faster Than We Are Eliminating Them?" ibid., 14 (July 1954): 14-18.

30. Franklin D. Richards, "The Mortgage Market Lenders Forgot," *The Mortgage Banker* 10 (December 1949): 7-9.

31. George W. Snowden, "This Is the Problem of Minority Housing," *The Mortgage Banker* 15 (October 1954): 20-22.

32. Everet Kinkaid, "Legal Restrictions Which Maintain Value," *Review of the Society of Residential Appraisers* 6 (April 1940): 3-8, and (May 1940): 13-14; H. Gordon Bollman, "Values in Blighted Areas," ibid., 8 (February 1942): 11-14; Frank E. Wilson, "Blighted Areas," ibid., 8 (April 1942): 11-12; Albert E. Dickens, "Measuring Blight," ibid., 10 (May 1944): 3-8.

33. "Neighborhood Rating," *Review of the Society of Residential Appraisers* 6 (August 1940): 7-9; Arthur A. May, "Let's Look at Appraisals," ibid., 6 (June 1940): 2-5.

34. Leslie H. Bamburg, "Values Guarded by Owners Alliance," *Review of the Society of Residential Appraisers* 10 (June 1944): 14-15; Oscar I. Stern, "Long Range Effect of Colored Occupancy," ibid., 12 (January 1946): 4-7; George L. Schmutz, "Sidelights on Appraisal Methods," ibid., 14 (June 1948): 18; George A. Phillips, "Racial Infiltration," ibid., 16 (February 1950): 7-9.

35. Talmadge D. Auble, "Residential Appraisals in the Postwar Period," *Appraisal Journal* 12 (January 1944): 47-48; Arthur A. May, "Appraising the Home," ibid., 19 (January 1951): 18-25.

36. Charles Abrams, "The New 'Gresham's Law of Neighborhoods': Fact or Fiction," *Appraisal Journal* 19 (July 1951): 324-37.

CHAPTER 5

FAMILY VALUES, MUNICIPAL ZONING, AND AFRICAN AMERICAN FAMILY LIFE

MARSHA RITZDORF

Concurrent with the passage of the 1964 Civil Rights Act, Daniel Patrick Moynihan, secretary of labor in the Johnson administration, sent to his boss what was at that time an internal memorandum. It was released in 1965 as a Labor Department report, "The Negro Family: The Case for Action." In the report he described Black families as "a tangle of pathology." Specifically, he referred to the natural (or moral) rightness of the patriarchal family. "Ours is a society which presumes male leadership in private and public affairs." He specifically listed female-headed households as pathological on the face, with no reference to any characteristics other than the sex of the head of the family unit. He characterized such households as part of the "Black Matriarchy." He further stated, inaccurately, that "The white family has achieved a high degree of stability and is maintaining that stability."[1]

The inaccuracy of that statement is evident in U.S. Census statistics that show, for example, that in 1968 (the first year in which comparative data is available for African Americans) the divorce rate for Black women (10 percent) and White women (9 percent) was essentially the same.

75

While the proportion of children living in single-parent households was significantly higher in the Black community, the number of children living in White single-parent households accounted for two thirds of all children living in single-parent homes (a proportion that is still the same in 1993).[2] In addition, high numbers of employed mothers and high rates of child abuse, divorce, and domestic violence are all squarely rooted in the "idyllic White family" and were pervasive, albeit less publicized, in the 1960s as well.

Nonetheless, Moynihan's chilling portrait of the evil Black matriarch has been an enduring one for both the general public and policymakers, despite nearly thirty years of research and writing refuting the belief that female-headed households and/or Black families themselves are intrinsically deviant. This portrait is rooted in a traditional sociology of the family that

> has been noted for its absence of a strong tradition of theory and for being heavily normative, moralistic, and mingled with social policy and the social objectives of various action groups. Nowhere is this tendency more apparent than in its treatment of racial ethnic families in the United States.[3]

Today, while presented in far more subtle language, those same assumptions are embedded in policy solutions promoted across the American political spectrum. Yet this view is representative of a cultural myopia concerning what a family was, is, and can be in the future. In her 1995 book, legal scholar Patricia J. Williams discusses the ways in which a call for "family values" became the vehicle for attacking mothers and children. She unmasks the myths that sustain racism and provides a discursive analysis of White Americans' use of symbolism to justify perpetuating racially based social policy.[4]

While discussions of urban planning policy almost never explicitly center on the concept of family, it is implicitly central to the way our cities are planned and zoned. Unfortunately, the few cases in which the concept of family is discussed point to planning's acceptance of prevalent myths. In 1989, for example, Peter Hall subtly but clearly accepts and supports an analysis that blames Black families for their own impoverishment. Citing a limited set of investigators from the post-Moynihan period, he accepts their assertions about the value of only a limited

So transportation policies perfect tool Then which to examine.

construct of a stable family without any reference to the larger body of scholarship on the survival strategies of the African American family. His work does not acknowledge the strong and stabilizing influences of extended families and fictive kin on poor Black neighborhoods that numerous researchers have reported.[5] He goes out of his way to make a case that the Kerner Commission report, which clearly labels "white, institutionalized racism" as the causal factor of the riots of the 1960s, is a distortion of reality.[6] He ignores research, contemporary with his writing, that addresses the diversity of family life. Yet given his fame in the field of planning, Hall's writing is very influential.

One of the most serious problems with an analysis that assumes a patriarchal family will solve all families' problems is a denial of the facts of contemporary family life for people of all races and ethnicities. For example, rapid social changes are affecting all families. While more White women become poor, single parents through divorce (with many of these women leaving their husbands because of domestic violence or abuse of their children) and more Black women become single parents, having already been poor, whether married or single (two out of three), the resulting poverty is the same.[7]

At the heart of these debates is the ethnocentric view that assumes a nuclear family consisting of a husband, wife (who stays at home in the best possible case), and one or more children as the only acceptable norm for United States society. Because the definition of what a family is has great symbolic significance, it is currently at the center of an array of larger political battlegrounds. The debate about the context and consequences of new forms of family escalates as the racial and ethnic composition of the United States undergoes dramatic change. More and more communities contain pluralities of different racial and ethnic groups. In other communities, the resident White population doggedly looks for ways to barricade themselves against newcomers who are different, whether Black, Asian, Latino or Native American.[8]

The debate over family form is one that demands a political judgment about the fate of one particular conception of the family and family life. "What is in fact at stake is a certain idealized form of the nineteenth century middle class family: a male dominated nuclear family that both sentimentalized childhood and motherhood and, at the same time, celebrated domestic life as a utopian retreat from the harsh realities of industrial society."[9] In this perfect family, mothers stay home and fathers go to work. While only 10 percent of American families overall conform to this norm in the 1990s, a much higher proportion of White children (75 percent) than Black children (26 percent) do live in a home that contains two parents, although in most of those homes the mother works outside of the home.[10] Therefore, the nuclear family norm becomes a convenient fulcrum for public policy that punishes women and children who by choice or by chance live outside its confines. As a result, policies that ignore the realities of change in American family life are far more punitive as they affect the African American community. The question of what forms contemporary African American families would choose— if their incomes and access to good schools, good housing, and employment were equalized with that of White families—simply remains unanswered.

However, counter to the prevailing mythology, history shows that from the postslavery era to the 1960s, the majority of Black families were two-parent households (70 percent of which were nuclear families).[11] Yet it is true that the Black family developed in a different form than that which is held up as the White, middle-class ideal. Urban poverty, underemployment, and unemployment led to a higher proportion of both female-headed and other alternative household forms for African Americans. The viability of a nuclear household supported by one male breadwinner was already problematic when job opportunities diminished for Black men and women more severely than for White families during the depressions of the 1870s and 1890s. "The exclusion of African Americans from skilled trades and factory work led to poverty and unemployment that made it necessary for many families to pool their resources and for others to split up, as members went different directions in search of work or security. . . . Married Black women were five times more likely to work for wages than were married White women."[12] A large and growing number of scholars of Black family life do not always agree on everything, but they concur that "black Americans' cultural orientation encourages family patterns that are instrumental in combating the oppressive racial conditions of American society."[13] These patterns include the nuclear family only as part of a much broader construct of family including real and fictive kin beyond the bonds of mother, father and children. Collectivity, it is argued, is a prominent characteristic of African American culture.[14]

PLANNING, PUBLIC POLICY,
AND THE AFRICAN AMERICAN FAMILY

Governmental policymakers, including planning officials, most often rely on only the one, narrow view of the African American family as represented by Moynihan. One need go no further than current and mean-spirited welfare reform for an example. The policies being suggested will eliminate the right to ensure that all Americans have a federally protected safety net and will fall most heavily on women and children, especially African American women and their children. The allocation of public resources to families and children in poverty should not be decided based on a stereotype of family life.

Herbert Gutman, preeminent contemporary historian of the Black family, identifies the work of nineteenth-century historian Philip Bruce as the important link between the popular view of African Americans as "unworthy" and the pseudoscientific data of Social Darwinism (survival of the fittest) that dominated U.S. Progressive Era thought up through the end of World War I. Bruce's scholarship is significant because his train of thought shows how policymakers justified the beginnings of the vicious patterns of residential segregation that exist to this day.[15] George Fredrickson explains the relevance of these theories and their impact on public policy in *The Black Image in the White Mind*. He explains that it was assumed that Blacks were so inferior that they would not even survive for many more generations. Therefore, the major social concern became how to "control" them so that they would not blend into the larger, White population during the period of their decline and disappearance.[16] Segregationist policies such as exclusionary zoning and restrictive residential covenants were a logical outgrowth of such thought.

These policies ensured that no matter how "middle class" Black families became, in both social and economic terms, they were closed out of the opportunities available to other Americans. For example, ethnic White Americans, often because of extraordinary efforts to save money at the expense of their health and the comforts of family life, were able to purchase their own homes in the post–World War I housing boom, although the homes were in separate neighborhoods from those of "native-born Whites."[17] This opportunity was not open for Black families. In this era, "the most striking feature of Black life was not slum conditions, but the barriers that middle-class Blacks encountered in

Kenneth Jackson, CRABGRASS
FRONTIER

trying to escape the ghetto."[18] The home ownership programs—which were developed under the aegis of the Home Owners Loan Corporation (which itself started during the Great Depression) and which were a major financer of urban housing through the post-World War II period— actually exacerbated the problem. Not only were the new programs rooted in bigotry in regards to the rating and appraising of urban neighborhoods occupied by ethnics and Blacks, but "the pattern of discrimination was continued until at least 1970 by the Federal Home Loan Bank board, whose examiners routinely 'red-lined' postal zip codes in which the symptoms of racial change and falling values were observed."[19] Not only were African Americans unable to obtain the funds to improve their own homes and neighborhoods in the city, but the Federal Housing Administration developed a program to guarantee home loans for new housing (which in many cases made it cheaper to own than rent) that denied access to the Black family. Bias in favor of homogeneous White suburbs was the fundamental principle of FHA loan guarantees. Their underwriting manual endorsed restrictive zoning and was uncompromising in its prejudices against "inharmonious racial or nationality groups" (see Part 5.B, this volume).

Black families, no matter how economically solvent, were thus unable to get help either to renovate their own neighborhoods or to move into the suburbs with the millions of White Americans who were fleeing the city with federal help. However, the rich extended-family life of kin and community available to most African Americans helped sustain them through the hardest of times. Studies of nineteenth-century and early-twentieth-century cities revealed much tighter kinship ties for African Americans than for Whites.[20] While there were slightly higher rates of female-headed and extended households in the Black community, the difference was a source of strength and support. In the 1950s and 1960s, for example, researchers in many regions of the country demonstrated that alternative family forms in Black communities were flexible, effective ways of coping with long-term poverty and growing unemployment rates among men that exceeded those of the depressions of the late 1800s.[21]

Multiple studies challenge the mythology with which White America burdens Black families. These studies show (among other myth-defying results) that the average crack addict is a White middle-class male in his forties; that, while a much higher proportion of White men in poor neighborhoods actually married their partners when they reported themselves pregnant, most Black fathers provided some degree of support; and importantly, as William Julius Wilson and other scholars assert, that lack of decent jobs, not a lack of interest, is often the roadblock to marriage. Indeed, a national study revealed that Black, unmarried fathers had more contact with their children and gave their mothers more informal support than did absent White middle-class fathers.[22] Income studies reveal that regardless of the political rhetoric that insists (as did Vice President Dan Quayle in 1992) that "Marriage is the best anti-poverty program we have," two out of every three poor Blacks living in single-parent families were poor before their family status changed.[23] Also lost is the fact that poor Black families in this nation live their lives in segregated (often hypersegregated) inner-city neighborhoods with little job opportunity and little or no access to transportation to get them to jobs in suburbs for which they might qualify. These same suburbs use the mechanisms of planning and zoning to make these families unwelcome as residents.

LOCAL PLANNING AND ZONING

Although Black families had been excluded from certain residential neighborhoods for generations, by the use of tools that included racial zoning (see Silver, Chapter 2, this volume), it was the massive building of American suburbs after World War II that showed the determination to use governmental policies and monies to create a society clearly segregated by both economic class and race. "What was unusual in the new circumstances was not the presence of discrimination—Jews and Catholics as well as Blacks had been excluded from certain neighborhoods for generations—but the thoroughness of the physical separation that it entailed."[24] For example, in 1960, not one of the 82,000 residents of Levittown on Long Island, New York, was Black.[25]

While in theory the original zoning ordinances of the late teens and early 1920s were designed to protect all citizens against the intrusion of noxious commercial and industrial uses into their neighborhoods and to preserve property values, they were in actuality devices designed to keep poor people, as well as industry, out of affluent areas.[26] In addition, loan regulations for the Federal Housing Authority (FHA) and

Veterans Administration (VA) demanded rigid racial discrimination (see planning-related documents in Part 5.B, this volume, for selections from the FHA underwriting manual). By the 1950s, rising fears about the civil rights movement and the concurrent explosion of affordable suburban housing meant that it was no longer the privilege of only the very affluent to buffer themselves against "blight" (a euphemism for the poor and Blacks); this opportunity for self-segregation now expanded across the broad range of the White middle class.[27] Both the mass migration of Blacks to American cities between 1940 and 1960 and the 1954 decision to outlaw school segregation (*Brown v. Board of Education*) added to the momentum.[28] But as eminent urban historian Kenneth Jackson notes, it was a not often discussed fear that fueled the suburban exodus: "pervasive fear of racial integration and its two presumed fellow travelers—interracial violence and interracial sex."[29]

While the fear of violence is discussed in the planning literature, White fear of interracial sex is not, except in rare cases. Yet this fear played, and still plays, a significant role in increasing segregation. The persistent untruths about Black men's uncontrollable sexuality and the "loose morals" of Black women create a racial stereotype that brands them as unacceptable neighbors to Whites who assume, whether consciously or unconsciously, that these stereotypes are true. "For most, if not almost all, critics of the Black family, there is always at the back of the mind this myth, this image of Black America as Babylon."[30]

Studies such as Denton and Massey's *American Apartheid* prove that residential segregation by race is an insidious and growing problem. Regardless of such regulatory action as the Federal Fair Housing Act of 1968 and its revisions in 1988, "traditional family values" remain the bedrock on which communities can legally build better barriers to an integrated future. While many middle-class Black families are achieving better housing in suburban areas, the homes are primarily in majority African American suburbs.

The evolution and continued use of family definitions in municipal zoning ordinances during the late 1960s and 1970s and the legal discourse surrounding this issue are discussed elsewhere in this book (see Ritzdorf, Chapter 3, this volume). Looking at the history of the discourse about family in municipal zoning, one can see the "family ethic" quite clearly, although this tendency is perhaps less clear to the reader whose life experience is that of Whiteness or middle-class income or who does not feel "oppressed" by current societal structures. However, both racial prejudice and sexism drive the use of these supposedly neutral ways in which family is defined for the purposes of municipal zoning. The historical use of family definitions in zoning ordinances is to discriminate against those who, by choice or chance, live in family arrangements that do not conform to the traditional White middle-class nuclear family model.

My earlier work investigated this phenomena and its impacts in a variety of ways, including analyses of jurisprudence relating to the definition of family in zoning law and actual empirical studies of contemporary ordinances. I created a longitudinal database that, unfortunately, shows that even within the last ten years such definitions have gotten more, rather than less rigid in many communities.[31] Yet in 1991, a U.S. Census report revealed that one out of two or 32.3 million American children lived in a nontraditional family that contained people other than the biological parents and their offspring. Eight million lived in an extended family containing, for example, a grandparent, aunt, or uncle. Thirty percent of all children of single parents lived in an extended situation. Only 26 percent of all African American children live in a traditional nuclear family. Nonnuclear families were, and are, more likely to occur among disenfranchised groups, the impoverished, and those from different races and cultures.[32]

Zoning is consistently used to prevent the spatial extension of people of color into White, middle-class America. Stereotypes influence zoning regulations that attempt to keep alternative living arrangements out of neighborhoods called "single-family," where the mythical nuclear family resides. Perin states in her 1988 study of the relationship between land use and social order that Americans find the very presence of those of different status in their neighborhood to be an unsettling experience and are especially discomforted by female-headed and minority-headed households.[33]

Municipalities have the right (in all but three states—Michigan, New Jersey, and California) to establish zoning ordinances that determine the number of unrelated individuals who may share a household and whether they are to be considered a family. The typical ordinance defines a family as an *unlimited* number of individuals related by blood, adoption, or marriage but only a limited number of unrelated individuals living together as a single housekeeping unit. In some communities, no unrelated individuals are allowed to live together.[34]

Almost all American zoning ordinances contain a definition of family and these definitions have a history of use that is as old as zoning itself. Restrictive ordinances are still judicially supported in the 1990s. In a 1991 case (*Dinan v. Town of Stratford*), the Connecticut Supreme Court upheld a locality's restrictive family definition that allowed only a maximum of two individuals unrelated to the family of the occupant to live in a single-family unit. Although the American Planning Association, the first time taking an organizational stand against such restrictive definitions, filed an amicus curiae brief, the court praised traditional family districts and vacated the lower court decision that had invalidated the regulation since it regulated the user (the people) and not the use.[35]

While it is often argued that family definitions are irrelevant because communities never enforce them, the surveys I conducted in both 1984 and 1994 found that not to be true. In 1984, 91 percent of the communities defined family in their ordinance, and in 1994 that increased slightly to 93 percent. Furthermore, 60 percent in 1984 and 64 percent in 1994 established a numerical limit to the number of unrelated people who could live together as a family group. While 40 percent of the communities in both years they were surveyed defined family broadly as a "single housekeeping unit," the other 60 percent all established numerical definitions. Almost all communities separately regulate group quarters such as fraternity houses, so there is no reason to assume that the family definition is needed for that purpose.

In addition, the majority of ordinances set limits of six or fewer unrelated people while giving broad discretion for interpretation of who is related to whom. For example, two single mothers with two children each and one grandchild wishing to share a five-bedroom home could easily be prevented from doing so if the ordinance allows only three people unrelated to each other to live together. In 1984, 55 percent of the respondents said they had been asked to investigate complaints related to the violation of local family definitions, and in 1994, 54 percent responded that they had been asked to investigate complaints.

More important, these definitions are enforced. In 1984, 42 percent of the communities had enforced their family definition and required non-traditional family groups to change their lifestyle or location. In 1994, 47 percent of the communities responded that they had enforced their family definition since 1984.[36] In addition to family definitions, a community can use numerous other land use tools—such as large lot zoning, lack of transit, lack of zoning that allows child care or elder care, lack of zoning to allow accessory apartments or multifamily housing, and so on—to enforce a particular socioeconomic vision of family life. These options express the power of the community to discriminate as it sees fit within the limits of the law. It is only too easy for those who are disenfranchised by these decisions to recognize them for what they are: active, but legal, tools of racial, class and gender discrimination.

Planners can and should argue that implicit in the granting of land use and zoning powers to U.S. municipalities is a mandate to use those powers in a socially responsible way. The language of municipal zoning ordinances, like any culturally bound discourse, is a language that both persuades and informs us about societal values. Poor families, especially African Americans, are most affected by these definitions that regulate the intimate composition of households making it often impossible, for example, for two single-parent families (or an unmarried couple and their children) to share a single-family home.

While other tools are used together with discriminatory lending and realty practices to effectively segregate Black and White communities, family definitions are a more insidious and often overlooked zoning power, since they never mention anything to do with race and seem to address only familial relationships. However, with the high incidence of nonnuclear families, family definitions are another roadblock to free choice in housing for members of the African American community. The family types most affected by these definitions are single parents who wish to share their home with another single parent and their children; single parents sharing a home with a partner (including a child's father) to whom they are not legally married; any circle of fictive kin regardless of age or sex (for example, a single parent, her children, and an elderly friend); any family of choice made up of unrelated individuals and, possibly, in some states, extended families. In Kent, Washington, for example, the zoning ordinance considers a family member only as related "up to four degrees of consanguinity," which means a first cousin.[37]

All families' lives, regardless of race, ethnicity, or class, would be enhanced by residential neighborhoods that allow them the opportunity for true integration, not just of people but of options in the way they work and play out their daily home lives. The freedom to work at home, to have their children (or parents) watched at small neighborhood-based day care centers, to share living spaces with the companions of their choice, and to use the spaces within their homes as they choose (as long as they do not violate the safety and choice of others), should be the birthright of all citizens. Unfortunately, segregation of both people and uses remains the norm and will do so in the near future. The more isolated White Americans' residential environment from people of color and those with a different socioeconomic status, the "better" they perceive the neighborhood to be.[38]

A combination of class, race, and gender determines how affected a family is likely to be by the use of zoning as a social moral arbiter. That communities use narrow Eurocentric models to legitimize only certain family forms is a less obvious and, therefore, more insidious form of discrimination than the myriad other ways in which we validate race and class discrimination, and it is clearly a roadblock to free housing choice for millions of African American families and their children. While there are other, bigger problems for us to address in order to tackle issues of racism and impoverishment in the United States today, I would argue that the same rhetoric examined here and applied to zoning permeates almost all policy-making. It will be an enormous task to deconstruct and redirect the conversation if and when the political will to do so exists in this nation.

For most Americans, a family is the unit from which they draw their nurturance and sustenance. It is not a particular form, nor is it a simple, symbolic image. Indeed, researchers find that many people living in voluntary families (not related by blood, kinship, or marriage) are doing so to create for themselves a sustaining life unit. Extended families have been the lifeblood of millions of African American families for innumerable centuries. Absolutely no evidence exists that extended families are less nurturing or more immoral than the attenuated nuclear model that we cling to so dearly in the United States. Ample evidence does exist that the nuclear family as the ideal was indeed a product of industrialization; if one reads the political rhetoric carefully, the main concern about nonnuclear families is almost always expressed in economic terms. Hence, in his now infamous speech to the California Commonwealth Club after the 1992 uprisings in Los Angeles, then–Vice President Dan Quayle talked extensively of the economic impoverishment of single-parent households and how a "welfare check is not a father."[39] While it certainly is not a father, it provides more income than an unemployed father; furthermore, if one looks carefully at welfare history, it becomes clear that until very recently, any father, unemployed or not, but living in the home, doomed the children in most states to a denial of an AFDC claim. The father's presence therefore caused the legal (and often emotional) breakup of many homes.[40]

As discussed earlier, collectivity is a marked characteristic of African American culture. Poverty is, unfortunately, also a characteristic of the lives of a high proportion of African American families in the contemporary United States. So is the consistent presence of racial prejudice. Under these conditions, the extended family (whether constructed of blood relatives or fictive kin) is a pragmatic strategy for survival. The antiquated and narrowly constructed viewpoint that the nuclear family is the only acceptable norm must be reexamined in every aspect of our social policy, including municipal land use and zoning. "For African Americans, especially, the nuclear family model tends to offer an inadequate survival or pragmatic nurturing strategy. Instead, for African Americans, extended families increase the chances for improving one's situation."[41]

The concept of family should be a unifying rather than a dividing one in the United States. The development of planning policies that build on the strengths of a plurality of family forms will be necessary as we continue to undergo demographic and economic changes that affect our racial/ethnic composition, the structure of the labor market, and changing gender roles. Only by extending our vision to a diversified structure of family life will the United States move toward an egalitarian future and away from the continuing and escalating racial tensions in our communities.

NOTES

1. Daniel Patrick Moynihan, *The Negro Family: The Case For Action* (Washington, DC: United States Department of Labor, 1965).

2. This was calculated from data available on-line from the U.S. Bureau of the Census; charts used were titled but unnumbered. They are: "Living Arrangements of Children Under 18 Years Old: 1960 to Present," "Living Arrangements of Black Children Under 18 Years of Age: 1960 to Present, "Living Arrangements of White Children Under the Age of 18: 1960 to Present," and "Marital Status of the Population 15 Years and Older by Sex and Race: 1950 to Present." Compiled March 31, 1996.

3. Maxine Baca Zinn, "Family, Feminism and Race in America," *Gender and Society* 4.1 (1990): 71.

4. Patricia J. Williams, *The Rooster's Egg: On the Persistence of Prejudice* (Cambridge, MA: Harvard University Press, 1995).

5. Peter Hall, *Cities of Tomorrow: An Intellectual History of Urban Planning and Design in the Twentieth Century* (New York: Blackwell, 1988). For work on fictive kin, see, for example, Carol Stack, *All Our Kin: Strategies for Survival in a Black Community* (New York: Harper and Row, 1974).

6. National Advisory Commission on Civil Disorders (Kerner Commission), *Report* (Washington, DC: Dutton, 1968).

7. Women's Research and Educational Institute, *American Women: A Status Report* (New York: W. W. Norton, 1995).

8. Maxine Baca Zinn, "Family, Feminism and Race in America," 73.

9. James Davison Hunter, *Culture Wars: The Struggle to Define America* (New York: Basic Books, 1991), 180.

10. United States Bureau of the Census. See charts referred to in note 2 of this chapter.

11. Herbert G. Gutman, *The Black Family in Slavery and Freedom* (New York: Pantheon Books, 1976), 500.

12. Teresa L. Amott and Julie A. Matthaei, *Race, Gender and Work: A Multicultural Economic History of Women in the United States* (Boston: South End Press, 1991), 166.

13. See, for example, Harriet Pipes McAdoo, "Family Values and Outcomes for Children," *Journal of Negro Education* 60.1 (1991): 361-64; Andrew Billingsley, *Climbing Jacob's Ladder: The Enduring Legacy of African American Families* (New York: Touchstone, 1992); Jacqueline Jones, *Labor of Love, Labor of Sorrow: Black Women, Work and the Family from Slavery to the Present* (New York: Vintage Books, 1995); and Carol Stack, *All Our Kin: Strategies for Survival in a Black Community* (New York: Harper and Row, 1974).

14. Algea O. Harrison, M. N. Wilson, C. J. Pine, S. Q. Chan, and R. Buriel, "Family Ecologies of Ethnic Minority Children," *Child Development* 61.2 (1990): 347-62, and Stack, *All Our Kin*.

15. Herbert G. Gutman, "Persistent Myths About the Afro-American Family," *Journal of Interdisciplinary History* 6 (1975): 188.

16. George Frederickson, *The Black Image in the White Mind: The Debate on Afro-American Character and Destiny, 1817-1914* (Middletown, CT: Wesleyan University Press, 1987).

17. *Native-born White* is the term that the White, Christian middle class, whose ancestors were primarily Dutch, English, Scotch, and German and had arrived in the earliest waves of immigration, used to describe themselves around the turn of the century.

18. Kenneth Jackson, *The Crabgrass Frontier: The Suburbanization of the United States* (New York: Oxford University Press, 1985), 203.

19. Ibid.

20. See works cited in note 13.

21. See ibid. and Amott and Matthaei, *Race, Gender and Work*.

22. Stephanie Coontz, *The Way We Never Were: American Families and the Nostalgia Trap* (New York: Basic Books, 1992); William Julius Wilson, *The Truly Disadvantaged: The Inner City, the Underclass and Public Policy* (Chicago: University of Chicago Press, 1990).

23. Daniel Quayle, speech to Commonwealth Club, San Francisco, May 19, 1992; Wilson, *The Truly Disadvantaged*; Paula Ries and Anne J. Stone, eds., *The American Woman: A Status Report* (New York: W. W. Norton, 1992).

24. Kenneth Jackson, *The Crabgrass Frontier*, 241.

25. Ibid., 240.

26. Ibid.; Charles M. Harr and Jerold S. Kayden, *Zoning and the American Dream: Promises Still to Keep* (Chicago: Planners Press, 1989).

27. Richard Babcock, *The Zoning Game* (Madison: University of Wisconsin Press, 1966); Kenneth Jackson, *The Crabgrass Frontier*; Douglas Massey and Nancy Denton, *American Apartheid: Segregation and the Making of the Underclass* (Cambridge, MA: Harvard University Press, 1993).

28. Nicholas Lemann, *The Promised Land: The Great Black Migration and How It Changed America* (New York: Knopf, 1991); *Brown v. Board of Education*, 349 U.S. 294.

29. Jackson, *The Crabgrass Frontier*, 290.

30. Lerone Bennett Jr., "The Ten Biggest Myths About the Black Family," *Ebony*, August 1986, 123-24.

31. Marsha Ritzdorf, "Zoning and the Changing Family: Resistance or Change?," unpublished manuscript, 1995; idem, "Planning and the Intergenerational Community: Balancing the Needs of the Young and the Old in American Communities," *Journal of Urban Affairs* 9.1 (1987): 79-89.

32. U.S. Commerce Department, "The Diverse Living Arrangements of Children," Report no. P70-38 (Washington, DC: U.S. Government Printing Office, 1992).

33. Constance Perin, *Belonging in America: Reading Between the Lines* (Madison, WI: University of Wisconsin Press, 1988); Constance Perin, *Everything in Its Place: Land Use and Social Order in America* (Princeton, NJ: Princeton University Press, 1977).

34. In New Jersey, New York, and California, the state supreme courts, interpreting the state constitutions, have refused to use *Belle Terre* as the guiding case. Instead, they assert that in their state constitutions the rights of unrelated individuals are protected to the same degree (in regards to zoning jurisdiction) as are the rights of related individuals. See: *City of Santa Barbara v. Adamson*, 610 P. 2d 436 (Cal., 1980), *State of New Jersey v. Baker*, 405 A. 2d 368 (N.J., 1979), *McMinn v. Town of Oyster Bay*, 488 N.E. 2d 1240 (N.Y., 1985). In Missouri, on the other hand, the state supreme court upheld an ordinance that allowed only married couples to live together (no unrelated people to live together). See *City of LaDue v. Horn*, 720 S.W. 2d 745 (Mo. App., 1986).

35. *Dinan v. Town of Stratford Board of Zoning Appeals*, 595 A. 2d 864 (Conn., 1991).

36. Marsha Ritzdorf, "Zoning and the Changing Family."

37. Marsha Ritzdorf, "The Impact of Family Definitions on American Municipal Zoning Ordinances" (Ph.D. diss., University of Washington, 1983).

38. Marsha Ritzdorf, "A Feminist Analysis of Gender and Residential Zoning in the United States," in *Women and the Environment*, ed. Arza Churchman and Irwin Altman (New York: Plenum Press, 1994); Constance Perin, *Everything in its Place* and *Belonging in America*.

39. Daniel Quayle, speech to the Commonwealth Club, San Francisco, May 19, 1992.

40. Mimi Abramovitz, *Regulating the Lives of Women: Social Welfare Policy from Colonial Times to the Present* (Boston: South End Press, 1988).

41. Donna Yvette Ford and William L. Turner, "The Extended African American Family: A Pragmatic Strategy that Blunts the Blade of Injustice," *Urban League Review* 14.2 (1990-91): 80.

PART TWO

PLANNING
AND PUBLIC POLICY

See Part 5.B, C, D, G, I, and K, for documents related to planning and public policy.

THE PERSISTENCE OF RACIAL ISOLATION

The Role of Government
Action and Inaction

YALE RABIN

Since World War II, massive decentralization of White population, housing, and economic activity has transformed American metropolitan areas. Fewer than one-third of metropolitan area Whites now live in the central cities, and in the last four decades, more than two of every three new jobs in metropolitan areas have been located in the suburbs. Despite the significant numbers of middle-income Blacks who have found housing in the suburbs, low-income Blacks remain concentrated in central cities, where they continue to increase as a proportion of the population, and where they remain isolated from decentralizing metropolitan housing, from suburban employment opportunities, from decent schools, and from reasonable access to basic goods and services.

During the 1950s and 1960s, racial discrimination and its economically and socially disabling consequences became issues of national concern, recognized as the legacy of over a century of systematic de jure and de facto denial of equal opportunity to Blacks. Congress responded by enacting a series of sweeping civil rights laws. But not until the late 1960s was racial isolation recognized as a distinct form of residential segregation. Today, after three decades of effective White resistance to the enforcement of civil rights laws, that earlier recognition has changed into the widespread misperception that the persistent isolation and poverty of central-city Blacks are mainly the result of poor job skills, apathy, and dysfunctional behavior on the part of ghetto residents.

By skillfully manipulating and exploiting the racial and economic anxiety of Whites, cynical politicians have convinced an all too willing public that conditions that are in reality the effects of discrimination and isolation are instead the cause of those conditions and that past efforts to ameliorate those conditions have, at best, done no more than waste scarce public resources. Despite this revisionist rhetoric, the underlying reality remains unchanged.

Actually, a diverse array of government programs, policies, and actions have sustained and reinforced the segregated isolation of low-income central-city Blacks. If Congress succeeds in its current efforts to dismantle what little remains of any federal programs that assist or protect low-income families or individuals, the existing concentrations of social and economic distress will become permanent elements of our urban centers.

Despite this deplorable prospect, and the fact that race relations continue to be of major national concern, the issue of racial isolation is strikingly absent from the current political agenda. What follows provides an overview of the changes that have led to the current state of affairs. The focus is on the principal government actions that have led to racial isolation, the changes in government policy that have taken place in response to racial isolation, and the implications of those changes.

THE WEB OF
ISOLATING INFLUENCES

While overt racially discriminatory motives have often influenced public policy decisions—particularly at the local level—some of the most far-reaching racially discriminatory effects have been caused by policies and programs that are ostensibly racially neutral. Most prominent among these are federal and state transportation policies. These have interacted with more explicitly discriminatory actions such as local government land use regulation, the segregative policies and practices of housing and redevelopment authorities, the failure to enforce civil rights laws, several major decisions by the U.S. Supreme Court, and severe reductions in federal funding for education, housing, health care, and other social programs.

These are among the principal strands in the restrictive web of government action and inaction that have reinforced pervasive discriminatory attitudes. These diverse policies and activities differ widely in the nature and intensity of their impacts on the isolation of low-income Blacks. All reinforce the persistently widespread array of private market discriminatory practices—such as racial steering, mortgage redlining, and insurance redlining—which aggravate isolation.

Most of the public actions considered here have locational implications. Some exert a powerful influence on the spatial distribution of development, some influence the nature of development, while others establish conditions of access to the benefits of development. In considering the isolating effects of government actions, it is important to understand that some of the same policies that restrict the locational choices of low-income Blacks also disproportionately increase locational choices for Whites and provide incentives for them to move, thereby intensifying the racially isolating effects of government actions.

The process of metropolitan decentralization, which began early in the century and has accelerated since World War II, has been the inevitable consequence of pent-up demand and pre-existing trends. However, the spatial pattern of that decentralization has been shaped by the federal investment of tens of billions of dollars in new highways during the thirty-five years following the war. The resulting metropolitan highway networks have made millions of acres of suburban land accessible for development and continue to be instrumental in determining the spatial configuration of today's dispersed, auto-dominant, and auto-dependent pattern of suburban sprawl.[1]

As the White population and manufacturing, wholesaling, and retail establishments have left the central cities, the distribution of population, land use, employment opportunities, and amenities have all been drastically altered. The racially and economically polarizing changes set in motion by the interaction of these public policy-induced trends have become elements in a self-reinforcing process. The departure of middle-income Whites has reduced central-city tax revenues, thus undermining the fiscal ability of many cities to support their schools and to maintain previous levels of other municipal services. Black pupils have become the majority in many of these underfunded city public school systems; White merchants have abandoned low-income Black neighborhoods to follow the greater purchasing power of their previous White customers to the suburbs. And so the process grinds inexorably on.

The shift of housing and economic activity to dispersed locations on the new regional highway networks has drastically reduced ridership on, and undermined the viability of, existing urban public transit systems. By 1970, public transit patronage had fallen by over 70 percent from pre-World War II levels. Transit systems responded by severely curtailing service and greatly increasing fares. These changes drastically reduced the number of jobs accessible by public transportation. As a result, the ability of metropolitan area residents to satisfy basic needs and share in the diverse and expanding opportunities of the metropolitan area has become dependent on access to an automobile.

For low-income central-city Blacks, access to the suburban opportunities made possible by federal highway funds has been severely limited by a rate of public transit dependence more than twice that of their low-income White counterparts.[2] Compounding the systemically adverse impacts on low-income Blacks of the highway-dominant federal transportation program have been the widespread, intentionally discriminatory practices in the implementation of the highway program at the local level. Routes selected have frequently created barriers between Black and White neighborhoods; right of way acquisitions have disproportionately disrupted Black communities; and relocation practices have reinforced patterns of racial segregation.[3]

Further reinforcing these constraints on spatial mobility have been the formally segregative housing loan guarantee policies of the Federal Housing Administration (FHA), the exclusionary zoning practices of many all-White suburban municipalities, the concentration of public and assisted housing in central-city ghettos, the segregative tenant-assignment practices of local housing authorities, and the preferences given to local (mostly White) residents by suburban jurisdictions in the distribution of housing vouchers and certificates.

FEDERAL GOVERNMENT INTERVENTION

This process of increasing racial isolation was well under way by the early 1960s, when the need for remedial intervention by government to undo a century of officially sanctioned racial injustice achieved widespread public acceptance. Racial zoning and racially restrictive covenants had been struck down earlier by the U.S. Supreme Court, and in 1954 the Court had issued its landmark *Brown v. Board of Education* decision, declaring de jure school segregation to be an impermissible denial of equal protection.

In November of 1962 President Kennedy, by executive order, mandated an end to racial discrimination in all housing owned, operated, or assisted by the federal government. Beginning in 1963, Congress followed with a series of civil rights laws prohibiting racial discrimination in the benefits of federal programs, public accommodation, voting, employment, and housing.

However, the promise of these measures has never been realized, because local White resistance has been so strong and federal enforcement efforts have been so weak. Eleven years after the passage of the Fair Housing Act, the U.S. Commission on Civil Rights found that "victims of discrimination and segregation in housing have been largely unprotected by the federal government," and that "HUD [the U.S. Department of Housing and Urban Development] and the Department of Justice have failed substantially in their roles in administering and enforcing Title VIII of the Civil Rights Act of 1968." [4] The most important advances in civil rights that have been achieved in the courts, to date, have been the result of private litigation undertaken by nonprofit public interest law firms,[5] not federal government enforcement efforts.

From the mid 1960s to the early 1970s, Black impatience and frustration exploded into a series of violent and destructive disturbances that raged through the ghettos of many cities. These led to intensive national inquiries into the causes of these eruptions, inquiries that identified high levels of unemployment as one of the principal underlying factors. These findings made some outside observers more aware of the effects of ghetto isolation and lent a sense of urgency to the need to remove externally created barriers to the spatial mobility of low-income Blacks. However, for the opponents of the civil rights struggle, the riots provided the political ammunition needed to ultimately gain the upper hand. For those most hostile to the civil rights struggle, the riots provided the impetus to push for a steady shift to the right in the mood of the public and of those they elected to public office.

The McCone Commission first brought the issue of ghetto isolation to public attention in its report on the 1965 Los Angeles riots. That report suggested that Black unrest arose, in part, from lack of adequate transportation to jobs. HUD responded by initiating an experimental demonstration project that provided ghetto-to-suburb bus services in more than a dozen cities.

There followed several years of intensive studies, by John Kain[6] and others, that explored the relationship between ghetto isolation and high levels of Black unemployment. These studies accepted the existence of spatial isolation and inadequate public transportation as givens. Aside from disparities in transportation access, the only other isolating government action acknowledged in a few of the studies was exclusionary zoning. None of the studies questioned the fact that low-income central-city Blacks were spatially isolated from suburban employment. The focus of the studies was on the *significance* of lack of access to suburban employment, among other factors, as a cause of high levels of Black unemployment.

Based on their findings, these studies fell generally into two groups. Those in the first group concluded that racial discrimination among employers and lack of job skills among ghetto residents were more significant factors than lack of access;[7] those in the second group concluded that lack of access to suburban jobs was more than offset by the availability of unskilled jobs in the central cities.[8] These studies confirmed the existence of central-city racial isolation but did not address the problem of overcoming that isolation because they did not find it to be of primary significance.

In these circumstances, it is hardly surprising that it was not until 1974 that Congress explicitly addressed the growing problem of isolation. By that time, it was already no longer politically acceptable to express its concern explicitly in terms of race. As a consequence, the Housing and Community Development Act of 1974 included the rather clumsily expressed goal of "reducing the isolation of income groups within communities and geographic areas and the promotion of an increase in the diversity and vitality of neighborhoods through the spatial deconcentration of housing opportunities for persons of lower income."[9]

By the mid 1970s, large-scale racial violence in the cities had subsided. Although the experimental transit programs had assisted many central-city residents in finding employment, HUD, claiming that none of those transit projects had become self-supporting, withdrew its funding and terminated the projects. As the violence subsided, interest in the issue of spatial isolation declined substantially among policymakers and most academics. The fact that several more recent studies[10] have attributed far more importance to spatial isolation than did the early studies has had no effect on public policy. In retrospect it is clear that the conclusion that isolation was not a significant factor in Black unemployment was the product of a tragically narrow definition of the problem.

THE SUPREME COURT
AND ISOLATION

The first major decision by the Supreme Court since 1917[11] to address housing discrimination was *Shelley v. Kraemer* in 1948.[12] That decision [outlawed the widespread practice of including racially restrictive covenants in residential real estate transactions.]However, while decisions by the Court in response to litigation attacking school segregation, voting rights restrictions, and employment discrimination produced dramatic improvements in those areas, litigation challenging housing discrimination and residential segregation, even when successful, has had no significant impact on the ongoing process of isolating low-income Blacks in central-city ghettos.

Two important decisions by the Court during the mid-1970s raised substantial new obstacles to any future efforts to obtain relief in the courts from the public policies that exacerbated metropolitan decentralization and racial isolation. The first was the Court's rejection of metropolitan school desegregation remedies in *Milliken v. Bradley* (1974).[13] That decision both reinforced the isolation of central-city Blacks and provided a powerful incentive to White out-migration from the central cities by creating a legal sanctuary from integrated schools for White families in the suburbs of metropolitan America. In addition, that decision lent impetus to the tendency of many suburban jurisdictions to adopt exclusionary zoning and restrictive cost-inflating subdivision regulations.

The second decision, *Washington v. Davis* in 1976,[14] put an end to almost all efforts to obtain relief in the courts from those ostensibly race-neutral government policies and programs whose foreseeable effects severely and disproportionately disadvantaged Blacks and other low-income minorities. In that case, the Court ruled that only those government actions that a court determines to be *intentionally* discriminatory on the basis of race violate the Equal Protection clause of the Fourteenth Amendment to the Constitution. The Court declared that statistical evidence of discriminatory effect, which had been accepted by most district and circuit courts until then, was no longer adequate.

The principal activities affected by this ruling are transportation planning, local land use regulation, and the provision of municipal facilities and services. Absent "smoking gun" evidence of intent, not even data showing that a municipality had provided paved streets and street lights in all-White residential neighborhoods, and none of these facilities in Black residential neighborhoods, would be adequate to demonstrate a violation of the Equal Protection clause. This decision legitimized the consistent failure—or refusal—by government decision makers to recognize and consider the readily foreseeable, racially discriminatory effects of public policies and actions that have no explicitly racial aspects.

These and subsequent decisions by an increasingly conservative U.S. Supreme Court have seriously eroded earlier civil rights gains. They have narrowed the definition of discrimination and restricted access to the courts by civil rights plaintiffs. As a result, ostensibly racially neutral government policies and actions—whose *effects* have disproportionately disadvantaged low-income blacks by severely limiting their spatial mobility—continue to exert their isolating influence.

With the exception of a brief interlude during the Carter administration, the conservative mood of the country has continued to spread since the early 1970s. With one or two notable exceptions, each appointment to the U.S. Supreme Court since the beginning of the Nixon administration has moved the outlook of that body farther and farther to the right. The recent appointments by President Clinton of Justices Breyer and Ginsburg have stemmed, but not reversed, that trend. And during that period, the nature of the public response to the persistent problem of concentrated racial isolation has shifted as dramatically as has the metropolitan landscape.

THE TRANSFORMATION OF CIVIL RIGHTS POLICY

What were earlier seen to be effects are now claimed to be causes; increasing public indifference and hostility have replaced understanding and concern. Earlier concerns about externally imposed restrictions on spatial mobility have yielded to a growing tendency to attribute Black poverty and deplorable ghetto conditions to antisocial attitudes and behavior of ghetto residents.

This view derives considerable support from the prominent attention that the media gives to the violent events that punctuate daily life in the ghetto. Unfortunately, the generalizations formed from these negative images provide an expedient, but inaccurate, explanation of black poverty. They lend credence to racist stereotypes and divert attention from the legitimate struggle for survival of most ghetto residents.

Using these stereotypes, cynical and opportunistic politicians have promoted a national backlash against civil rights by raising and exploiting racial anxieties among White voters. They have skillfully employed race-coded images to portray Blacks as undeserving dependents on public welfare and threats to public safety, to discredit those perceived to be sympathetic to civil rights, and to characterize those who support social programs as advocates of Black interests. These techniques have become powerful and effective influences on the outcomes of elections at every level of government.[15]

It is ironic that these misperceptions are also reinforced, although perhaps inadvertently, by the recent proliferation of "underclass" studies, focusing, as they do, on the dysfunctional behavioral characteristics used to identify "underclass" membership. However, these characteristics—low male labor force participation and high welfare dependence, school dropout rates, and numbers of female-headed families—are found among fewer than half of all central-city Blacks living below the poverty level.[16]

Since 1980, it is this backlash that has formed the basis of federal civil rights policy. That policy now rests on several flawed assumptions: that the civil rights enactments of the 1960s fulfilled all government obligations to remove previously erected racial barriers to equal opportunity; that a remedy for racial discrimination should be available only to an individual who can demonstrate that he or she has been the direct victim of an intentionally racially discriminatory offense; that equal opportunity is a zero-sum game in which any effort to achieve equal rights for Blacks in an affirmative manner discriminates against Whites; that any remaining disparities in the status and condition of Blacks and Whites are attributable to the inability of Blacks to compete effectively in a presumably racially neutral marketplace.

By restricting the Black poor in central cities from access to growing suburban opportunities in employment, housing, and education, government actions have reinforced widespread private discrimination and perpetuated the burdens of a long history of racial oppression. The interactive effects of these government actions have subverted the intentions of Congress as expressed in a decade of civil rights legislation between 1963 and 1974. The failure of government, at all levels, to implement and enforce constitutionally mandated and legislatively enacted measures to protect civil rights is all too evident in the concentrated poverty and social disorder of central-city ghettos.

RESULTS OF POLICIES

As Whites and middle-income Blacks have left the central cities, the concentration and isolation of low-income Blacks in the inner cities have increased. By 1993, nearly 70 percent of all metropolitan-area Whites and almost 35 percent of metropolitan-area Blacks lived outside the central cities. Between 1970 and 1993, the number of Blacks living in central cities increased by 40.6 percent,[17] but during the same period the number of central-city Blacks living in poverty increased by over 108 percent[18] (see Table 6.1). From 26.0 percent in 1970, the proportion of central-city Blacks living below the poverty level increased to 35.8 percent in 1993.[19]

The failure of HUD and the U.S. Department of Justice to enforce antidiscrimination laws and regulations and to implement legislatively mandated deconcentration policies has compounded the racial isolation and economic distress of low-income central-city Blacks. This has led to the continuing concentration of federally subsidized affordable housing available to low-income Black families in the deteriorating, problem-ridden ghettos of the central cities.[20] As a result, 80 percent of all Blacks in public housing or in federally subsidized, privately owned project-based housing lived in the central cities in 1990.[21] Much of this housing supply is in severely deteriorated condition.

Table 6.1 Poverty Status of the Black Population in Metropolitan Areas

	1970	1980	1990	1993
U.S. Black population	22,580,000	26,397,000	30,836,990	33,040,000
Percentage Black	12.7	11.7	12.4	12.7
Percentage below poverty	33.5	32.5	31.9	33.1
In metro areas	16,771,000	20,505,000	25,568,000	28,398,000
Percentage of all Blacks	74.3	77.7	82.9	86.0
In central cities	13,140,000	14,957,000	17,367,000	18,478,000
Percentage of all Blacks	58.2	56.7	56.3	55.9
Percentage of metro Blacks	78.3	72.9	67.9	65.1
Below poverty	3,176,000	4,831,000	5,870,000	6,615,000
Percentage below poverty	26.0	32.3	33.8	35.8

SOURCES: U.S. Bureau of the Census, *The Social and Economic Status of the Black Population in the United States: 1973; Summary Population and Housing Characteristics, U.S.: 1990; Characteristics of the Black Population, U.S. Summary: 1990;* Center on Budget and Policy Priorities, "1989 Poverty Tables," based on data from the U.S. Bureau of the Census and other sources; "1993 Poverty and Income Trends," March 1995.

However, these deplorable conditions are far from uniformly distributed. In 1990, while 56.3 percent of all Blacks lived in central cities, more than two out of every five central-city Blacks (40.4 percent) lived in one of only twelve cities: New York, Chicago, Detroit, Philadelphia, Los Angeles, Houston, Baltimore, Washington, Memphis, New Orleans, Atlanta, and Cleveland.[22] It is in these cities, plus a few smaller majority Black cities, that the major concentrations of Black poverty are located.

A few effective efforts, such as the Gautreaux program in Chicago, have enabled low-income Black families to obtain housing outside socially and economically distressed inner-city neighborhoods. In addition, the Clinton administration, in 1994, launched a small experimental Moving to Opportunity (MTO) program in five cities.[23] However, because of the extremely limited scope of these efforts, they have had no discernible effect on the concentrations of race and poverty in the inner cities in which they operate. At one point in the frenzy of budget cutting for social programs, Congress eliminated future funding for MTO.

Until 1995, Gautreaux and MTO, employing HUD Section 8 certificates and vouchers, offered a limited potential for enabling some deconcentration. Those limited opportunities have been virtually eliminated by steady reductions in appropriations for federally assisted low-income housing. Appropriations for HUD-subsidized housing programs have been reduced by 82 percent since 1981, the beginning of the Reagan administration.[24]

Although outlays for assisted housing have remained relatively high, the gap between supply and need has widened dramatically during the 1970s and 1980s. In 1970, there were 300,000 more low-rent units than there were low-income renters. By 1991, there were 4.5 million *fewer* low-rent units than low-income renters.

Several reasons explain this change: (1) the number of low-income families in need of assisted housing has continued to rise; (2) the existing stock of privately owned low-rent housing—most of which is in the inner-city ghettos—has deteriorated and declined sharply; and (3) federal housing assistance, particularly in the provision of new certificates and vouchers, has fallen drastically. In 1977, HUD issued 247,000 new certificates and vouchers. By 1990, the number had declined to 70,000;[25] the congressional budget proposal for 1996 provides for only 30,000 additional certificates or vouchers. By removing the funding for MTO, this proposal also eliminates the already severely limited opportunities to use certificates or vouchers as a means of escaping the concentrated distress of inner-city ghettos.

PROSPECTS FOR THE FUTURE

The elimination of resources will continue to profoundly affect conditions within isolated central-city ghettos. The federal government has severely curtailed funding for the rehabilitation of existing public housing and has sharply reduced other programs that directly affect the quality of life of ghetto residents. The Center on Budget and Policy Priorities has tabulated the changes that have taken place in some of the relevant programs (see Table 6.2).

In the face of these funding losses, and the apparently firm control over Congress held by a socially and fiscally reactionary majority, future prospects for residents of the ghetto appear to be as hopeless as the grim reality of the present. Trapped by a congruence of isolating forces, low-income Blacks have been consigned to deteriorating, violence-ridden central-city ghettos, while White population and economic opportunities continue to drain away to surrounding suburbs and beyond.

Table 6.2 Reductions in Funding for Low-Income Discretionary Programs: 1981-1995

Employment and training	– $7.8B	– 58.8%
Community development block grants	– 1.8B	– 29.2%
Energy assistance	– 1.7B	– 53.8%
Legal services	– 140M	– 25.9%
Community services block grants	– 419M	– 47.4%

SOURCE: Center on Budget and Policy Priorities, "Funding for Low-Income Programs in FY 1994" (Washington, DC: Center on Budget and Policy Priorities, 1995), 6.

The physical deterioration and social disintegration that have resulted from this seemingly inexorable process have blurred the distinction between cause and effect.

The barriers to the employment of low-income Blacks have been reinforced, not only by the civil rights backlash but also by structural changes in the economy and by the recent large influx of Southeast Asian and Central American immigrants. In the shift from a goods-producing economy to an information services economy, the largest numbers of manufacturing jobs have been lost in those cities in which low-income Blacks are most heavily concentrated.

What remains for the unskilled are menial jobs, and these are being increasingly filled by the grateful and more compliant beneficiaries of our immigration policies, who are often willing to work under substandard conditions for substandard wages. Further compounding the distressed condition of low-income central-city Blacks have been the slowdown in overall employment growth and the current and pervasive wave of "downsizing" throughout the economy. As a result, millions of central-city Blacks have, in effect, been declared surplus to the needs of the national economy.

It appears that nothing short of the reindustrialization of the United States, or the undertaking of a massive program of public works by the federal government, would have the potential for reintegrating this discarded ghetto labor force into the national economy. Clearly, both of these alternatives fly in the face of powerful current trends. Manufacturing operations have been moving "offshore" to lower labor cost locations in Latin America and Southeast Asia for decades; and the last time the government undertook a massive, employment-generating program of public works, the millions who were out of work were overwhelmingly White.

Despite the growth in academic attention to the characteristics and condition of the underclass and the widespread media attention to drugs and violence in the ghetto, the issues of ghetto isolation and Black poverty are strikingly absent from the public agenda. What began as a blatant racial backlash now claims added legitimacy from a tidal wave of fiscal conservatism. So sweeping has been the political shift to the right in the Congress that support for any legislation intended to remedy, or even to mitigate, the effects of racial isolation seems, at present, inconceivable. As a consequence, the quality of life of those consigned to ghetto poverty is now determined, not by the nation's professed standards of equity and decency but by the expedient and inequitable outcomes of an increasingly unconstrained private market.

If this dismal view of conditions and events bears a reasonable resemblance to present reality—and the evidence strongly suggests that it does—then it is vitally important that these issues be brought into the open, into the arena of public political discourse. Without a doubt, there have been, and continue to be, deplorable responses by many ghetto residents to deplorable conditions, but, without a doubt, government bears an equally heavy responsibility for the constraints and conditions that have evoked these responses.

Only with widespread recognition of this responsibility and with acknowledgment that the healing process begun during the 1960s was interrupted, not completed, can there be any hope of a response by government that will be remedial, rather than repressive. The public and elected officials alike must confront the implications of the present course: that by returning, on the threshold of the twenty-first century, to the political, social, and economic values of the nineteenth, they will perpetuate the poverty and despair of over 6.5 million central-city Blacks.

NOTES

1. See, for example, Thomas M. Stanback and Richard V. Knight, *Suburbanization and the City* (Montclair, NJ: Allenheld, 1976); President's National Urban Policy Report, *A New Partnership to Conquer America's Communities* (Washington, DC: U.S. Government Printing

Office, 1978); Peter O. Muller, *Contemporary Suburban America* (Englewood, NJ: Prentice Hall, 1981); and Kenneth T. Jackson, *Crabgrass Frontier: The Suburbanization of the United States* (New York: Oxford University Press, 1985).

2. Yale Rabin, "Metropolitan Decentralization, Transit Dependence, and the Employment Isolation of Central City Black Workers," paper presented to the symposium "The Role of Housing Mobility in Achieving Equal Opportunity for Minorities," Urban Institute, Washington, DC, April 21-22, 1988.

3. Yale Rabin, "The Roots of Segregation in the Eighties: The Role of Local Government Actions," in *Divided Neighborhoods: Changing Patterns of Racial Segregation,* ed. Gary A. Tobin (Beverly Hills, CA: Sage, 1987), 208-26; idem, "Federal Urban Transportation Policy and the Highway Planning Process in Metropolitan Areas," *The Annals of the American Academy of Political and Social Science* 451 (September 1980): 21-35; idem, "Highways as a Barrier to Equal Access," *The Annals of the American Academy of Political and Social Science* 407 (May 1973): 63-77.

4. U.S. Commission on Civil Rights, *The Federal Fair Housing Enforcement Effort,* Report to the President (Washington, DC: U.S. Government Printing Office, 1979).

5. The principal public interest law firms in the civil rights area at the national level are: NAACP Legal Defense and Education Fund, NAACP, The Lawyers Committee for Civil Rights Under Law, Mexican-American Legal Defense and Education Fund, National Lawyers Guild, National Committee Against Discrimination in Housing, and the ACLU. The landmark metropolitan desegregation case, *Gautreaux v. Chicago Housing Authority,* was brought by Business and Professional People in the Public Interest, a Chicago public interest organization.

6. John F. Kain, "Housing Segregation, Negro Employment, and Metropolitan Decentralization," *Quarterly Journal of Economics* (May 1968): 175-97; and John F. Kain, "Housing Segregation, Black Employment, and Metropolitan Decentralization: A Retrospective View," in *Patterns of Racial Discrimination,* ed. George von Furstenburg, Bennett Harrison, and Ann R. Horowitz (Lexington, MA: D. C. Heath, 1974), 5-20.

7. Thomas H. Floyd Jr., "Using Transportation to Alleviate Poverty: A Progress Report on Experiments under the Massachusetts Transportation Act," in *Conference on Transportation and Poverty,* ed. John F. Kain (Brookline, MA: American Academy of Arts and Sciences, 1968), 10; Joseph D. Mooney, "Housing Segregation, Negro Employment, and Metropolitan Segregation: An Alternative View," *Quarterly Journal of Economics* (May 1969): 299-311; John M. Goering, "Transporting the Unemployed," *Growth and Change* 2 (1971): 34-37; Bennett Harrison, "The Intrametropolitan Distribution of Minority Economic Welfare," *Journal of Regional Science* 12 (1972): 23-43; and Sanford H. Bederman and John S. Adams, "Job Accessibility and Underemployment," *Annals of the Association of American Geographers* (September 1974): 378-86.

8. Edward Kalachek and John Goering, "Transportation and Central City Unemployment" (Washington, DC: U.S. Department of Housing and Urban Development, 1970); Roger Noll, "Metropolitan Employment and Population Distribution and the Conditions of the Urban Poor," in *Financing the Metropolis,* ed. John P. Crecina (Beverly Hills, CA: Sage, 1970), 481-509; and Bennett Harrison, "Discrimination in Space: Suburbanization and Black Unemployment in Cities," in von Furstenburg et al., *Patterns of Racial Discrimination,* 21-53.

9. The Housing and Community Development Act of 1974, 42 U.S.C. § 5301(c)(6).

10. James L. Shanahan, "Impaired Access of Black Inner City Residents to the Decentralized Workplaces," *Journal of Economics and Business* 28.22 (1976): 156-60; Peter M. Hutchinson, "Transportation, Segregation, and Labor Force Participation of the Urban

Poor," *Growth and Change* 9.1 (1978): 31-37; Quentin Gilliard, "Reverse Commuting and the Inner City Low Income Problem," *Growth and Change* 10.3 (1979): 12-18; Yale Rabin, "The South Midtown Freeway and the Regional Transportation Planning Process: Effects on the Travel, Employment, and Housing Opportunities of Kansas City Missouri's Low Income Black Population," unpublished report to the U.S. District Court for the Western District of Missouri, 1979; John E. Farley, "Disproportionate Black and Hispanic Unemployment in U.S. Metropolitan Areas," *The American Journal of Economics and Sociology* 46.2 (1987): 129-50; and Mark A. Hughes, "The Formation of the Impacted Ghetto: Evidence From Large Metropolitan Areas, 1970-1980," *Urban Geography* 2.3 (1990): 265-84.

11. *Buchanan v. Worley*, which outlawed the practice of racial zoning in 1917.

12. *Shelley v. Kraemer*, 334 U.S. 1 (1948).

13. *Milliken v. Bradley* (Milliken I), 418 U.S. 717.

14. *Washington v. Davis*, 426 U.S. 229 (1976).

15. Thomas B. and Mary D. Edsall, "Race," *The Atlantic Monthly* 267.5 (May 1991): 53.

16. See, for example, William J. Wilson, *The Truly Disadvantaged: The Underclass and Public Policy* (Chicago: University of Chicago Press, 1987); Erol R. Rickets and Isabel V. Sawhill, "Defining and Measuring the Underclass," *Journal of Policy Analysis and Management* 7.2 (1988): 316-25; Ronald B. Mincy, "What Happened to the Underclass During the Economic Recovery?" testimony before the Joint Economic Committee of the U.S. Congress, May 25, 1989, *The Underclass* (Washington, DC: U.S. Government Printing Office, 1989); Paul A. Jargowsky and Mary J. Bane, "Ghetto Poverty: Basic Questions," in *Inner City Poverty in the United States*, ed. Lawrence E. Lynn Jr. and Michael G. H. McGeary (Washington, DC: National Academy Press, 1991), 16-67.

17. Bureau of the Census, *The Social and Economic Status of the Black Population in the United States: 1973*; and *Black Population in the U.S.*, Current Population Reports, March 1995.

18. Scott Barancik, *1989 Poverty Tables: Based on Data from the U.S. Bureau of the Census and Other Sources* (Washington, DC: Center on Budget and Policy Priorities, 1991).

16. Ibid.

17. Douglas S. Massey and Nancy A. Denton, *American Apartheid: Segregation and the Making of the Underclass* (Cambridge, MA: Harvard University Press, 1993), 188-216; see also John B. Kasarda, "Inner-City Concentrated Poverty and Neighborhood Distress: 1970-1990," *Housing Policy Debate* 4.3 (1993): 283.

18. Connie H. Casey, *Characteristics of HUD-Assisted Renters and Their Units in 1989* (Washington, DC: Office of Policy Development and Research, HUD, March 1992).

19. Bureau of the Census, *Black Population in the U.S.*

20. The MTO program, patterned after the Gautreaux program in Chicago, was enacted in 1992 during the Bush administration but was not implemented until the Clinton administration.

21. Center on Budget and Policy Priorities, "Funding for Low-Income Programs in FY 1994" (Washington, DC: Center on Budget and Policy Priorities, 1995), 6.

22. Derived from a tabulation by Cushing Dolbeare, consultant on housing and public policy, from HUD Budget Summaries FY 1976 through FY 1990.

7

URBAN PLANNING, EQUITY PLANNING, AND RACIAL JUSTICE

NORMAN KRUMHOLZ

E quity planning is a conscious attempt by some professional urban planners to devise and implement redistributive policies that move resources, political power, and participation toward low-income groups. Racial justice is an important priority for equity planners since deprivation and discrimination disproportionately affect low-income communities of color.

While traditional, or mainstream, urban planners focus on the physical environment, where they attempt to adjust public works and other investments in a way that is consistent with social and economic trends and sound principles of design, equity planners seek to reduce inequalities in cities and metropolitan regions. As a result, the practice of equity planners often differs from that of most urban planning professionals and sometimes operates in tension with the local political process. Traditional planners use a generally middle-class value system that emphasizes aesthetics, efficiency, and the value of real property, while trying to fulfill local objectives of physical well-being concerning both immediate needs and those of the foreseeable future. They are inclined to accept a technical role and let elected politicians or planning boards set goals and choose the ends of their work.

Equity planners challenge or reject this role. They assert that planning should aim to provide a better future for all city people, not just select groups. As such, planners must be concerned with means as well as ends. They believe that highest priority should be given to the most disadvantaged and troubled groups since existing societal institutions are inherently biased against the interests of those at the bottom of our social system. Equity planners also argue that the beautiful and efficient cities of which all urban planners dream cannot be realized so long as cities are dominated by racial segregation, poverty, and slums. In their view, the rebuilding of cities is much more than a matter of brick and mortar, no matter how splendid the architecture or expensive the structures; the city is people, and unless most of them are employed, respected, and self-respecting, the gleaming towers of downtown are a mockery.

HISTORIC ROOTS OF EQUITY PLANNING

Equity planning is not a new or isolated approach in urban planning. It is part of an alternative planning practice that, in the United States, goes back to the turn of the century. It includes the "material feminists" who sought communal kitchens and similar housing arrangements to liberate women from drudgery,[1] as well as designers like Alice Constance Austin who attempted to translate the cultural values of Southern California into a planned and egalitarian social landscape.[2] It also includes the work of Patrick Geddes, who drew plans based on a cooperative model of city evolution, and of the Regional Planning Association of America, which in the 1920s dedicated itself to social betterment through planning.[3] Planners, including Clarence Stein and Henry Wright, sought to reduce land coverage and housing density to achieve social and health objectives.[4] Lawrence Veiller in New York City worked to improve slum housing through more rigorous health and density codes, while Rexford Tugwell and the other planners of the Greenbelt Town Program worked toward healthier new planned communities in the New Deal.[5] At the same time, the New Deal Tennessee Valley Authority's regional projects helped provide jobs, electric power, and flood control to a badly depressed region. Then in the 1960s, the Model Cities program, the War on

Poverty, and planners such as Paul Davidoff, Herbert Gans, and Chester Hartman sought to aid minority, poor, and working-class urban populations victimized by urban renewal and other public policies.[6]

While these planners developed alternative approaches to planning practice, issues of racial justice were, and until recently continued to be, largely peripheral to mainstream planning practice. In his classic history *American City Planning Since 1890*,[7] Mel Scott barely mentions race or racism. T. J. Kent, in a thoughtful discussion of planning and politics, *The Urban General Plan*,[8] gives no discussion to issues of race. The two most important city plans of the 1960s, those of Philadelphia (1961) and Chicago (1965), do not deal with race as an issue, even though disadvantaged and racially segregated minority populations were growing sharply in both cities (see Part 5.D). Both plans place highest priority on maintaining and attracting the White middle class. In these and other important planning documents, thoughts of racial justice, redistribution, and participation were pushed to the margins of planning practice.

The civil rights ferment of the 1960s brought issues of race closer to the mainstream of planning thought and practice. More published articles on Black issues and by Black authors appeared in the *Journal of the American Institute of Planners* (JAIP).[9] In March 1969, an entire issue of *JAIP* was devoted to "The Cities, the Blacks, and the Poor."[10] The Great Society also produced planning programs focused on minorities and the poor, the most important of which were the Community Action Programs (1964) and the Model Cities Program (1966).[11] In 1971, the Code of Ethics of the American Institute of Planners was modified without opposition to reflect concern for poor and otherwise vulnerable populations (see Part 5.I). The significant change reads this way:

> A planner shall seek to expand choice and opportunity for all persons, recognizing a special responsibility to plan for the needs of disadvantaged groups and persons, and shall urge the alteration of policies, institutions, and decisions which militate against such objectives.

By the early 1990s, American Planning Association (APA) annual conferences included a substantial percentage of papers on social equity; between 1992 and 1994, 20 to 30 percent of all conference papers dealt with equity issues.[12] Also, the APA itself sponsored and published *Planning and Community Equity*, a collection of papers showing practicing planners how to address equity issues in their day-to-day work.[13] Although a number of scholarly books appeared in the 1990s based on planning and race,[14] this was the first one aimed specifically at planning practitioners and sponsored and printed by APA.

RACE AND URBAN PLANNING

Lest too optimistic a picture of professional planning be painted, it should be noted that in 1995, fewer than 10 percent of APA members were racial minorities and fewer than 4 percent were African Americans.[15] Yet African American and Latino populations are more and more likely to live in central cities where they make up larger and larger proportions of the entire population. In 1870, 80 percent of the African American population lived in the South in an agricultural setting; by 1970, 80 percent lived in the urban North.[16] Latinos, mostly of Puerto Rican, Cuban, and Mexican extraction, have also flooded into many American cities. As a result, since 1970, many American cities have dramatically changed their racial coloring.

Census data underscore the magnitude of that change. In the twenty largest cities of the Northeast and Midwest, the White population fell by over 2.5 million, or 13 percent, between 1960 and 1970 and another 4 million, or 24.3 percent, since 1970. During the same period, the African American population grew by 40 percent in these cities. Snowbelt cities have become important residential locations for African Americans. Latino groups are more likely to be concentrated in Sunbelt states that border on Mexico, but their numbers are rising in such Snowbelt cities as Chicago, New York, Hartford, and others. Racial and ethnic segregation of these minority populations—in all American cities, by lenders, realtors, developers, and landlords—is the rule.[17]

Among non-Latino Blacks in 1990, 62 percent lived in blocks that were 60 percent or more Black and 30 percent in neighborhoods that were 90 percent or more Black. Among Latinos, 40 percent lived in blocks that were 60 percent or more Latino. In most American cities, more than 70 percent of the population would have to move to achieve full integration.[18]

That cities have become the homes for people of color means that they have also become the residence for many of society's least fortunate. The evidence of severe poverty is overwhelming. In 1989, 13 percent of Americans were poor; they were disproportionately people of color highly concentrated in large cities.[19] This is not to suggest that all people of color are deprived. By the 1990s, more African Americans, with more resources than ever before, were respected members of the middle class, with an increasing number living in the suburbs. Their advance is noteworthy and heartening. Yet about a third of all African Americans were poor. Many, suffering from multiple disadvantages of residential isolation, racial discrimination, and poverty, were persistently poor. Commentators described them as "the underclass," a figure that has been estimated from 6 to 15 percent of the entire poverty population in the United States.[20]

Jobs, wealth, and economic opportunities have migrated outward in our metropolitan regions while poor minority communities in central cities have become increasingly isolated. Stark contrasts now exist between central cities and their surrounding suburbs.[21] In 1990, median income levels were almost 30 percent higher in the suburbs than in central cities. Poverty rates in central cities averaged 18 percent compared with 8 percent in suburbs.[22] In some cities, these disparities were even more extreme: New York City had a 1990 poverty rate of 19 percent, while its suburban poverty rate was 7 percent; in Detroit, the comparative figures were 30 percent in the city and 6 percent in the suburbs.

Income disparities also hold true in matters of race. The median family income for Black families has been only two-thirds that of White families for many years. Census statistics show this clearly. In 1970, the median income for White families was $10,236 and for Black families it was $6,279. In 1992, the White family median was $38,909 and the Black family median was $21,161. In 1966 (the first year comparative data are available for Black and White populations), 11 percent of the White population lived at or below the poverty line while 42 percent of Black Americans lived in poverty. In 1992, the number of Black Americans in poverty was still one-third of all Blacks. White poverty remained constant at slightly over 11 percent.[23]

The number of female-headed families with children—families that are particularly likely to live in poverty—has almost doubled since 1960. In 1992, 31 percent of Black households were single-parent and female-headed. Sixty-eight percent of children growing up in such households lived below the poverty level.[24] Young mothers and their children make up the fastest growing portion of America's homeless population, now estimated at one to three million.[25] Poverty and lack of education among parents are often associated with low birth weights, high rates of disease, and learning deficiencies, restricting these children in their later lives and imposing huge costs on society.

The combination of poverty and racial segregation contributes to very poor performance by children in the public school systems of the largest American cities (see Table 7.1). Disproportionate numbers of Black and Latino students currently do poorly on standardized tests in reading, writing, and mathematics. No modern society can develop its full economic, social, or intellectual potential with one-fourth of its children growing up under such conditions.[26]

Population projections make reform of these deplorable conditions especially urgent. Between 1995 and 2020, the U.S. Census projects a total U.S. population increase of 60 million, 79 percent of which will be among minority groups. With this growth, incomes and educational achievements among minority groups must improve radically or the average income and educational levels of all Americans will be drastically reduced.

Crime is also a major problem in the inner city. Different statistical sources paint somewhat different pictures, but on one question there is general agreement: Blacks in disproportionate numbers are both perpetrators and victims of crime. Blacks are about twice as likely as Whites to be victims of robberies and aggravated assault and six or seven times more likely to be victims of homicide. Homicide is the number one killer of young Black men, with a 1990 death rate about double that of American soldiers in World War II. The increase in the rate of urban violence is staggering: in New Haven in 1960, there were 6 murders, 4 rapes, and 16 robberies; in 1990 there were 31 murders, 168 rapes, and 1,784 robberies. In Milwaukee in 1965, there were 27 murders, 33 rapes, and 214 robberies; in 1990 there were 165 murders, 589 rapes, and 4,472 robberies. New York City had 244 murders in 1951; every year for the last decade the total was nearly 2,000. Crime is highly concentrated; ghettos are as dangerous as they seem and fear of violence helps shape their economic and demographic landscape.[27]

Clearly, the most extreme poverty in America, the highest unemployment rates, the greatest threat to personal safety are now found in

Table 7.1 Selected Statistics for 94 Large U.S. Cities 1960 to 1990 (in percentages)

	1960	1970	1980	1990
Population as percentage of U.S.	26.1	22.5	20.9	20.1
Percentage of minority population	18.9	24.1	37.1	40.1
Unemployment rate	5.5	4.7	7.3	8.1
Percentage employed in manufacturing	25.3	22.1	17.4	14.0
Median family income as percentage of U.S. median family income	106.7	100.4	92.6	87.5
Family poverty rate	17.2	11.0	13.6	15.1
Percentage of population in census tracts with more than 40% poverty	8.0	5.1	8.1	10.8
Female-headed families with own children as percentage of all families	7.9	10.4	13.8	14.5

SOURCE: Adapted from HUD, *Empowerment: A New Covenant with America's Communities*, July 1995, p. 14.

geographically isolated, economically depressed, and racially segregated inner cities and older declining suburbs. Inner cities have become warehouses of America's poorest citizens.[28]

WHY EQUITY PLANNING?

Traditional urban planning has not dealt effectively with these issues. Instead of giving high priority to disadvantaged populations and attempting to redistribute resources, power, and participation in their direction, city planners have tried, in as apolitical a way as possible, to serve "the community as a whole" with unitary land use plans depicting a "better" future. Much more is needed if planners are to have anything useful to say about poverty and racism. Indeed, if planners are ever to realize their goals of achieving the beautiful and efficient cities for which they plan, much higher priorities must be given to the slums and their deprived residents.

If the data presented in the first part of this chapter are accepted and linked to a personal value system that defines poverty, security, and racial justice in urban America as high priorities, then equity planning has a powerful claim on the entire urban planning profession. Society has a basic social responsibility to help preserve the conditions necessary for civilized life. Planners do as well. A powerful initial reason for urban planners to move in the direction of greater equity is to ensure that all children grow up with adequate nutrition, that civil order prevails, that public education prepares a person for a working life, and that all people can achieve a decent basic standard of living.

The second reason that equity planning has a powerful claim on the profession is one of political reality: Shifting populations of color have resulted in a rapid increase in the number of Black elected officials and a significant increase in the number of Black political jurisdictions. In 1970, there were only 1,469 Black elected officials nationally; by 1993, this figure had grown to 8,015.[29] This means that urban planners are more likely to be working in jurisdictions where a majority of the population is made up of people of color and working, as well, for Black or Latino elected superiors.

It is unlikely that these elected officials will be of one mind on the proper objectives of urban planning or public policy in general. After all, Carl B. Stokes of Cleveland did not govern the same way as Tom Bradley of Los Angeles; nor was the career of Harold Washington of Chicago similar in terms of style or content to that of Coleman Young of Detroit. But all of these African American mayors insisted on a higher measure of racial justice, and all wanted to do more for the constituencies that elected them. An equity-oriented planning practice fits nicely with these objectives.

Simple justice is a third reason that the pursuit of a more equitable, racially just society should appeal to urban planners and others. To be an effective principle, justice requires that we put ourselves in the shoes of another person and apply the reciprocal rule: "Do unto others as you would have them do unto you." If we assert our right to be treated fairly, we have a duty and an obligation to protect the rights of others. As the philosopher John Rawls makes clear, a just society is the kind of society that free, equal, and rational people would agree to establish in order to protect their own self-interest.[30]

Whether or not equity planners have ever heard of John Rawls, they try to adhere to the principles of a just society similar to those he sets out in his writing. They also try to pay close attention to the concerns for the disadvantaged cited earlier in the APA Code of Ethics. When public decisions have redistributive effects, they try to ensure that the gains go to those most in need of help. When sacrifices must be made, equity planners work to see that they are made by those most able to sustain them. The constraints on equity planning—working with groups having little influence or wealth and sometimes working against the grain of local business and politics—seem formidable. Yet in many cities, a core of equity planners, managers, and elected officials has found ways to advance a progressive agenda, advocate for the disadvantaged, increase racial justice, and stay in office. These kinds of municipal innovations are more widespread than is generally recognized and they represent a collection of innovative ideas and a powerful source of political and economic leverage.

EXAMPLES OF SUCCESSFUL EQUITY PLANNING

From 1969 to 1979, the Cleveland City Planning Commission pursued an equity planning agenda under three different mayors.[31] Their over-arching goal was to create "more choices for those who have few," shorthand for choosing up sides in favor of Cleveland's minority poor and working-class residents. The commission applied their goal to a wide series of public and private issues, including zoning, transportation, regional low-income housing, and neighborhood development. The planners made some waves in the process but also accomplished much good. One example of their approach has to do with transportation planning.

Most planning agencies define transportation problems in terms of rush hour, congestion, freeway access, or the need for more off-street parking. However, the commission's overarching goal led them to define Cleveland's most significant transportation problem in a different way; the need was to improve the mobility of Cleveland's transit-dependent population—those families without automobiles. In 1970, this was about 45 percent of all poor families in Cleveland. These transit-dependent families, the planners reasoned, had been disadvantaged by our national decision to opt for an automotive society. While those of us with cars enjoyed unrivaled mobility, the transit-dependent increasingly had less service at higher fares and could reach fewer destinations as new development decentralized the region. So given the equity-oriented perspective of the planning staff, the transit-dependent rider became the top priority of the planning staff.

By 1975, the staff had been involved in over five years of negotiations with state and county politicians and transit officials that ultimately led to the establishment of the Cleveland area Regional Transit Authority (RTA). During the negotiations, the planning staff argued that cheaper fares and expanded mobility for the transit-dependent population should be the transit system's highest priority. Everyone else argued in favor of costly new rail lines to distant suburbs and of people movers or other elaborate downtown distribution schemes. The planners' analyses made clear that fixed-rail systems would not provide much new mobility to the transit-dependent but would draw resources away from the more suitable bus services. Accordingly, the planning staff did everything they could to discredit rail proposals, including writing reports, op-ed articles, and analyses. They were successful. When agreement on the service package was finally reached, the planners had stopped rail extensions and won much of what they demanded:

1. A twenty-five cent fare would be initiated and maintained for at least three years.
2. Senior citizens and the handicapped would ride free during non-peak periods (twenty hours daily) and pay only half fare during the four peak hours.
3. Service frequencies and route coverage within the city would be improved.
4. A demand-responsive transit service for the elderly and handicapped—door-to-door and dial-a-ride—would be initiated and supported on a citywide basis.[32]

Another outstanding example of progressive politics and equity planning is found in Chicago during the 1980s. Planner Robert Mier and his associates helped build a coalition of White, African American, and Latino neighborhood groups that endorsed and helped elect the city's first Black mayor, Harold Washington. Washington took Mier and members of the coalition into city hall, where they challenged the interests of suburbs and downtown business by involving many who had been disenfranchised. Washington built solid support, particularly in communities of color and used it to build a progressive neighborhood-based program.

Washington also ran his campaign and then his office in an open, empowering manner. He sought to maximize diversity in the city's decision-making process, to open information channels, and to supply better information and better analyses to all citizens.

Three efforts were especially noteworthy. First, Mier and his fellow planners wrote the "Chicago Economic Development Plan, 1984," a model of progressive planning.[33] Its vision was driven by the notion that questions of economic equity—that is, who gets what kind of job and the resultant income—were inextricably bound to the practice of public urban economic development planning. The plan proposed to use the full weight of the city's leverage—tax incentives, public financing, city purchasing, infrastructure improvements—to generate jobs for Chicago residents. City resources were seen as public investments with a targeted return in the form of job opportunities for Chicagoans. Specific hiring targets for minority and female employment were set; 60 percent of the city's purchasing was directed to Chicago businesses; 25 percent of this would go to minority and women-owned firms.

The plan also sought to encourage a model of balanced growth between downtown and city neighborhoods. It offered public support to private developers to build projects in "strong" market areas if they were willing to assist neighborhood economic development projects in "weaker" market areas. Developers could support neighborhood-based community development corporations by providing them with technical or legal assistance, entering into joint ventures for neighborhood projects, or contributing to a low-income housing trust fund. As a final objective, the Plan drew up a regional, state, and federal legislative agenda to advance its interests.

Second, the city chose to "build on the basics" by working to transform the floundering local steel industry, by working with existing small manufacturers to maintain or increase employment and improve productivity, and by making demands on corporations already in place. Noting changes in the global economy, the city government worked to cut business deals directly with foreign corporations. They also worked to provide jobs for city residents and to direct municipal purchases toward city producers and suppliers. The city even sued the Playskool Corporation when it shut down a plant, forcing employees out of work in apparent violation of a contract involving an earlier city loan.

The planners also undertook a research project evaluating industrial displacement. They found that a major force driving businesses out of

Chicago was not high wages or taxes but incompatible real estate activity; gentrifying residential neighborhoods were expanding and encroaching on light manufacturing districts bordering the Chicago River, forcing many firms to move or close. To combat this trend and save jobs for local, mainly non-White residents, Chicago passed a new zoning ordinance to protect these industrial areas.

Third, the city worked intensely with neighborhood groups to turn the focus of public assistance toward the neighborhoods and away from subsidies to large, downtown businesses. Mier was able to build on his earlier founding of the Center for Urban Economic Development at the University of Illinois, which provides assistance to grassroots and neighborhood groups. In fights over the World's Fair, sports stadiums, street and highway investments, business taxes, and linkage fees, what was saved on the conventional subsidies to large businesses could be transferred into city budgets for housing, neighborhood improvement, public services, and the protection of existing jobs in small businesses throughout the city. Despite the opposition by downtown business, Mayor Washington's coalition held together and enlisted a grudging cooperation.

Jersey City offers the most recent example of equity planning, in the work of Rick Cohen, appointed Housing and Economic Development director in 1985.[34] On taking office, Cohen found that Jersey City had some major housing problems: 7,199 housing units stood vacant and uninhabitable, and the number of rental units had dropped from 63,156 to 58,110 between 1970 and 1980. Rents were rising much faster than the median income of residents, and speculators were driving renters out of their homes and "mothballing" the units for future conversion to condominium use. Clearly, Jersey City was being affected by the overheated New York City market just across the Hudson River.

Cohen introduced an anticondominium conversion bill and immediately obtained pledges from most of the big developers either to include affordable housing units in their projects or to donate to an affordable housing trust fund. The ten billion dollar Newport development on the Hudson River, which included several upscale apartment buildings, reserved 18 percent of the units at rents substantially below average for the use of low-income tenants.

Cohen believed that, with the right kind of urban planning, new developments could serve both the needs of the developers and upper-income tenants and the needs of the poor residents of Jersey City. Cohen believed training, job creation, and set-asides could provide many job opportunities for Jersey City residents, while linkage arrangements, which tie developments in "hot" areas to contributions for benefits in "cold" areas, would help provide affordable housing. Cohen also emphasized the need for racial diversity on his staff, raising the number of minority project managers from one to more than half of his staff. In addition, he stressed the need to work closely with neighborhood-based organizations.

Other cases in equity planning have been documented in many other cities.[35] Some examples include the following:

San Diego, where planners created a Housing Trust Fund for affordable housing funded from development fees[36]

Denver, where the city's Comprehensive Plan promotes scattered site public housing and regional racial integration[37]

Boston, where planners initiated a linkage policy, housing trust fund, and broad city support for neighborhood-based development corporations[38]

Portland, Oregon, where planners promoted downtown planning by citizen committees and freedom-of-information policies[39]

Santa Monica, California, where planners implemented a rent-control ordinance and a moratorium on large-scale development[40]

These cases make clear that equity planning can be done and can produce real benefits for poor communities of color.

HISTORIC STRATEGIES FOR SEEKING RACIAL JUSTICE

Historically, the equity planner's role has been, first of all, to defend and protect the interests of the clients, most often poor communities, usually of color. At a minimum, advocacy planners have identified how and when allocations move resources and power away from the poor and suggested, after analysis, alternative schemes in favor of this clientele. At a maximum, they have influenced the system on their clients' behalf. Mainstream planning concerns involving open space, upscale housing, or urban design have been, when necessary, subsumed to the clients' need for affordable housing, more and better security, low-cost transit services, jobs, the removal of racial barriers, and the like.

Second, the equity planner provides as wide a range of alternatives and opportunities as possible, leaving individuals free to define their own needs and priorities. Government efforts to alleviate poverty frequently emphasize a "service strategy," in which the government provides, or subsidizes the private provision of, particular goods and services. Unfortunately, these efforts have often failed to satisfy the needs of those whom they supposedly serve.

In the interest of maximizing choices, the equity planner supports expanded reliance on an "income strategy," and seeks to provide individuals with the means and the opportunity to obtain those goods and services that they perceive as best fulfilling their needs. One role equity planners play in improving choices for those who have few is to encourage racial and class integration in nonintegrated neighborhoods in the city and region. An example is the "Fair-Share Plan for Public Housing in Cuyahoga County" proposed by the Cleveland City Planning Commission in 1970. The purpose of the plan was to offer subsidized housing choices across the entire county for Cleveland's poor families.[41]

Third, the equity planner must recognize the crucial role played by legal, political, economic, and social institutions in promoting and sustaining inequities. Necessary changes will often require alteration in the laws, customs, and practices of these institutions. To the extent possible, the equity planner tries to reform these institutions while informing the client of their workings and the procedures necessary to achieve institutional changes. Examples of success in this area include weakening the real estate practices of blockbusting and racial "steering" for racial change, as was done in Shaker Heights, Ohio, and Oak Park, Illinois,[42] and helping to end redlining through the passage of laws such as the Community Reinvestment Act (1975) and Home Mortgage Disclosure Act (1976).[43] (See Part 5.K.)

The equity planner has to recognize the politics of his or her position on behalf of those less favored by urban conditions. Obviously, the less favored are not the most powerful nor, in many cases, the most numerous city constituency. The equity planner's recommendations are not accepted in all cases and equity planners always have to acknowledge the role of the interests of more favored individuals or groups. Good advocates demand that conflict should be understood, clearly articulated, and submitted to the relevant executive, legislative, or judicial body for resolution. Thus, the equity planner does not seek consensus but rather strives to identify, clarify, and publicize what may be opposing interests.

THE FUTURE OF EQUITY PLANNING AND RACIAL JUSTICE

Efforts to achieve racial justice are likely to be most effective in cities where people of color make up a majority or have elected powerful minority leadership. In that case, political support may exist for strongly targeted local redistributional policies. But there is a larger world that must come to share a concern for racial justice. To move that world, equity planners must develop a regional, state, and national agenda that raises the needs of the most deprived as a matter of justice and conscience but also as part of a broader, lower-middle-class agenda. Affordable housing for working- and lower-middle-class households, educational enrichment for children from lower-income families (who will make up an essential part of our nation's future workforce and protect our standard of living), investment in neighborhood (rather than downtown) redevelopment, raising the minimum wage, health benefits for all; these are the kinds of programs that positively affect a broad political spectrum of people while also helping the poor.

In the long run, the planners' most important contribution may be to remind our political and civic leadership that certain fundamental problems rooted in our society cannot be taken for granted, if we value the continued stability and prosperity of our nation.

NOTES

1. Dolores Hayden, *The Grand Domestic Revolution* (Cambridge, MA: MIT Press, 1981).

2. Dolores Hayden, *Seven American Utopias* (Cambridge, MA: MIT Press, 1976), 300-301. See also: Mike Davis, *City of Quartz* (New York: Vintage Books, 1992), 24-25.

3. Simon Eisner, Arthur Gallion, and Stanley Eisner, *The Urban Pattern*, 6th ed. (New York: Van Nostrand, 1993), 594.

4. Ibid., 168.

    5. Mel Scott, *American City Planning Since 1890* (Chicago: American Planning Association Press, 1969), 336-41.
    6. Norman Krumholz and Pierre Clavel, *Reinventing Cities: Equity Planners Tell Their Stories* (Philadelphia: Temple University Press, 1994), 13-14.
    7. Scott, *American City Planning Since 1890*.
    8. T. J. Kent, *The Urban General Plan* (Berkeley, CA: University of California Press, 1964).
    9. Robert A. Catlin, "The Planning Profession and Blacks in North America and the U.K.," paper presented to the Associated Collegiate Schools of Planning at Oxford, U.K., in July 1991. An exception to the general lack of early planning articles dealing with race and class is Melvin M. Webber's thoughtful article "Comprehensive Planning and Social Responsibility" in the *Journal of the American Institute of Planners* 30 (November 1963): 232-40.
    10. *Journal of the American Institute of Planners* 35 (March 1969).
    11. The Community Action Program aimed to combat poverty by directing federal resources to politically free-standing agencies involving "the maximum feasible participation" of the poor. Model Cities was intended to aid the poorest, most deteriorated city neighborhoods by coordinating grant programs within a metropolitan strategy.
    12. Telephone conversation with Frank So, associate director, American Planning Association, August 1995.
    13. American Planning Association, *Planning and Community Equity* (Chicago: APA Planners Press, 1994).
    14. Among the books relating to race and urban planning are: Pierre Clavel and Wim Wiewel, eds., *Harold Washington and the Neighborhoods, 1983-1987* (New Brunswick, NJ: Rutgers University Press, 1991); Robert Mier, *Social Justice and Local Development Policy* (Newbury Park, CA: Sage, 1993); Robert A. Catlin, *Politics and Urban Planning: Gary, Indiana 1980-1989* (Lexington, KY: University of Kentucky Press, 1993); William W. Goldsmith and Edward J. Blakely, *Separate Societies* (Philadelphia: Temple University Press, 1992).
    15. Internal membership data, American Planning Association, Chicago, 1995.
    16. Douglass S. Massey and Nancy A. Denton, *American Apartheid: Segregation and the Makings of the Underclass* (Cambridge, MA: Harvard University Press, 1993), chapter 2.
    17. Roderick J. Harrison and H. Weinberg, "Racial and Ethnic Segregation in 1990" (Washington, DC: U.S. Bureau of the Census, April 1992).
    18. Dan Gillmor and Stephen Doig, "Segregation Forever?" *American Demographics* 14 (January 1992): 48-51. See also Massey and Denton, *American Apartheid*.
    19. William Julius Wilson, *The Truly Disadvantaged: The Inner City, the Underclass, and Public Policy* (Chicago: University of Chicago Press, 1987).
    20. Richard Nathan, "Will the Underclass Always Be with Us?" *Society* 23 (March-April 1986): 57-62.
    21. William H. Frey and Elaine Fielding, "Changing Urban Populations: Regional Restructuring, Racial Polarization, and Poverty Concentration," *Cityscape* 1.2 (1995).
    22. Department of Housing and Urban Development, "Empowerment: A New Covenant with America's Communities" (Washington, DC: U.S. Department of Housing and Urban Development, 1995).
    23. U.S. Department of Commerce, United States Bureau of the Census, *Current Population Reports*, abstracted in *Statistical Abstract of the United States 1994* (Washington, DC: U.S. Bureau of the Census), tables 723-28, pp. 473-75.
    24. Ibid., tables 49 (p. 48) and 729 (476).

25. Philip L. Clay, "The Unhoused City," in *The Metropolis in Black and White,* ed. George Galster and Edward W. Hill (Philadelphia: Temple University Press, 1992), 95.

26. Anthony Downs, *New Visions for Metropolitan America* (Washington, DC: Brookings Institution, 1994).

27. Adam Walinsky, "The Crisis of Public Order," *Atlantic Monthly,* July 1995.

28. David Rusk, *Cities Without Suburbs* (Washington, DC: Woodrow Wilson Center Press, 1993). See also Gary Orfield and Carole Ashkinaze, *The Closing Door* (Chicago: University of Chicago Press, 1994); Martin Carnoy, *Faded Dreams: The Politics and Economics of Race in America* (Cambridge, MA: Harvard University Press, 1994).

29. Joint Center for Political Studies, "Black Elected Officials, 1993" (Washington, DC: Joint Center for Policy Studies, 1994).

30. John Rawls, *A Theory of Justice* (Cambridge, MA: Harvard University Press, 1971). See also Sue Hendler, *A Reader in Planning Ethics* (New Brunswick, NJ: Center for Urban Policy Studies, 1995).

31. "Cleveland Policy Planning Report" (Cleveland, OH: City Planning Commission, 1975).

32. Norman Krumholz and John Forester, *Making Equity Planning Work* (Philadelphia: Temple University Press, 1990), chapter 8.

33. Department of Economic Development, "Chicago Economic Development Plan, 1984" (Chicago: Department of Economic Development, 1984). See also Krumholz and Clavel, *Reinventing Cities,* chapter 4.

34. Krumholz and Clavel, *Reinventing Cities,* chapter 2.

35. Pierre Clavel, *Progressive Cities* (Philadelphia: Temple University Press, 1990); Krumholz and Clavel, *Reinventing Cities.*

36. Krumholz and Clavel, *Reinventing Cities,* chapter 10.

37. Ibid., chapter 7.

38. Ibid., chapter 6.

39. Ibid., chapter 5.

40. Ibid., chapter 9.

41. Ibid., chapter 6.

42. W. Dennis Keating, *The Suburban Racial Dilemma* (Philadelphia: Temple University Press, 1994).

43. The Community Reinvestment Act tries to ensure an adequate flow of credit to poor and working-class neighborhoods by providing a means to challenge lenders' practices; the Home Mortgage Disclosure Act mandates that lenders collect and display the location of all mortgages they make.

CHAPTER

8

GARY, INDIANA

Planning, Race, and Ethnicity

ROBERT A. CATLIN

By the mid-1990s, Gary, Indiana, had suffered the same fate as most of its Midwestern U.S. counterparts. Industrial giants from the early 1900s until the early 1960s, many cities witnessed decline due to post-World War II suburbanization, White flight, and the corresponding increase in the African American population. Beginning in the early 1960s, automation started to take its toll. Finally, by the mid-1970s, came the flight of high-paying, low-skilled jobs to the Sunbelt and/or overseas; in virtually all cases, this loss was not offset by new service industry employment. Examples of cities experiencing this phenomenon include not only Gary but also Detroit, Cleveland, Milwaukee, Boston, New Haven, Philadelphia, Pittsburgh, and to a certain extent, even New York City and Chicago.[1] What distinguishes Gary from these other cities is that revitalization was made more difficult because this city had no institutions bound to it by place, such as "Fortune 500" corporate headquarters, universities with major research capabilities, or large medical centers that could ensure community stabilization for, if no other reason, their own survival.[2]

On the other hand, Gary had the initial distinction of being a planned community. It was designed and built by U.S. Steel in 1903-1906 as a site for the corporation's production in the Midwest. Did starting as a "planned" city help to guide the city's growth and development in an orderly, rational manner? At what period of time were planning decisions most important? And finally, what role did the dynamics of planning and race play? Briefly, Gary's experience showed that (1) starting as a "planned city" did not guide the city's growth in an orderly and rational manner; (2) planning decisions, or actually nondecisions, were most crucial immediately after World War II; and (3) planning and race dynamics played a major role. This chapter attempts to answer these questions in greater detail.

GARY'S DEVELOPMENT: 1900-1951

The U.S. Steel Corporation established Gary in 1903. Needing a location midway between its major sources of iron, ore, and coal, U.S. Steel picked a barren, unpopulated site forty miles square and bordering Lake Michigan about thirty miles from downtown Chicago.[3] At first, the company intended to build only a steel mill, but because 10,000 workers were needed to staff this plant, housing was required within walking distance to streetcar lines. Therefore, U.S. Steel developed a full-scale city as well.[4] In April 1903, plans were finalized for the new plant and city with the municipality being named for Judge Elbert Gary, then U.S. Steel's board chairman.[5]

The town was laid out on an old gridiron pattern with no attention given to standard city planning techniques of that time such as wide diagonal streets radiating from central points, the integration of residential land with open spaces, and the protection of natural features such as lakes, rivers, and wooded areas. Actually, the period of 1890-1905 was ripe with *new* planning techniques such as Ebenezer Howard's garden city theory (1898), which placed built environments within "gardens" or permanent open space. The British town of Letchworth exemplified the garden city theory; its development began in 1903, three years before ground was broken for the City of Gary.[6] George Metzendorf's 1912 design for Margarethen-Hohe in Germany's steel-making Ruhr Valley placed housing, shopping, and schools separate from the mills in a garden city setting surrounded by permanent open space.[7] In contrast, Gary's town site was set behind the steel plant, which took up almost the entire lakefront.[8] To make matters worse, U.S. Steel built half a town. The first subdivision, Emerson, had wide streets, large lots, and complete utilities, including electric, gas, water, sewer, and telephone. It was designed as a residential area for the mill's executives and the city's professional class. Also accommodated in the more modest dwellings in the eastern part of Emerson were craftsmen, foremen, and skilled workers employed in the mill, and clerical workers and small shopkeepers employed in related retail trade and service activities. Speculators bought the portion south of the Ninth Avenue railroad tracks and created instant slapdash housing for unskilled laborers who were 90 percent of the mill's workforce. Within a few years, this area—unplanned and soon crowded, without paved roads, running water, or sewerage—turned into a festering slum.[9]

The first subdivision's western sector, with the city's most expensive dwellings, housed the upper middle class, almost exclusively White Anglo-Saxon Protestant (WASP) and Republican. Emerson's skilled workers were either native-born American Whites of English or Scotch-Irish extraction, or descendants of "old immigrant" Germans, Austrians, and Scandinavians. On the southside lived the unskilled workers that U.S. Steel recruited directly from eastern and southern Europe. Poles were the largest group, followed by Serbians, Croatians, Hungarians, Greeks, and Italians.[10] After World War I began and immigration from Europe ceased, African Americans coming from the rural South met the increased demand for unskilled labor. These new residents at first mixed with eastern and southern Europeans in the south side but by the end of the 1920s African Americans were concentrated in that neighborhood now known as the Central District. The former White residents relocated to the outlying communities of Glen Park and Tolleston.

As seen in Table 8.1, Gary grew quickly and spectacularly between 1910 and 1930. From dunes and barren marshlands in 1906, the city grew to 16,802 people in 1910, 55,378 in 1920, and 100,426 by 1930. Under the leadership of William A. Wirt, an innovative school system enrolled almost 5,000 children by 1913 and by the early 1920s had established a

Table 8.1 Gary, Indiana's Changing African American Population, 1900 to 1990

Year	Total Population	Black Population	Percentage of Change in Black Population from Previous Decade	Percentage of Blacks in Total Population
1910	16,802	383	—	2.3
1920	55,378	5,299	1,383.6	9.6
1930	100,426	17,922	238.2	17.9
1940	111,426	20,394	13.8	18.3
1950	133,911	39,123	92.3	29.2
1960	178,320	69,123	76.2	38.8
1970	175,415	92,695	34.1	52.8
1980	151,953	107,644	16.1	70.8
1990	116,646	93,982	−10.2	80.6

SOURCE: Statistics derived from selected annual decennial federal censuses.
NOTE: Gary not mentioned in 1900 census of places with populations of more than 2,500.

national reputation for preparing the children of immigrants to function in American society.[11]

The period from 1920 to 1929 was Gary's Golden Age. New banks, hotels, and office buildings arose in the downtown area, and the mill expanded its facilities. In 1926 alone, the city issued almost $2.5 billion in building permits (1995 dollars).[12] Public improvements for parks, utilities, and streets totaled a 1995 equivalent of well over $300 million between 1925 and 1930.[13] Despite these expenditures, Gary was able to keep its tax rate well below the average of all U.S. municipalities with populations between 25,000 and 100,000.[14] In addition to the steel mills and related industries, Gary became a state convention center as well. In 1928, eleven major statewide associations brought 12,000 visitors to Gary, with each spending a 1995 equivalent of $150 per day.[15] Urban planning, however, was not a factor in this growth. The city established a city plan commission in 1919 and it hired a consultant firm allied with Daniel Burnham to prepare a land use plan that featured large lakefront parks, boulevards, a civic center, and a proposed zoning ordinance.[16] However, U.S. Steel, the local real estate board, and the railroads fought the plan. U.S. Steel opposed a boulevard that was to run along Lake Michigan much like Lake Shore Drive does in nearby Chicago because it would have taken up room needed for future plant expansion. The mill superintendent stated, "The steel corporation has no apologies to offer for the way the city was first laid out and the accomplishments."[17] By 1923, the plan was dead, and the commission ceased to function because it repeatedly failed to secure a quorum needed to elect officers.[18]

Gary's population grew slightly during the depression years despite layoffs in the mill and related industries. Even as the depression lingered into the late 1930s Gary still gave outward appearances of a thriving progressive city. A Federal Writers Project guidebook praised Gary:

> Still adolescent, fresh, and vigorous, Gary with an estimated population of 115,000 (1938) has assured her place among the cities of the midwest as one of the leaders in industry, education, recreation, and architecture.[19]

This guidebook hardly mentioned Gary's African American community, which grew from only 383 residents in 1910 to 20,394 in 1940, 18 percent of the population. An in-depth study of Gary's African Americans conducted during that time would have found a population far more impoverished and decimated in spirit than their White counterparts, due to systematic, institutionalized racism. In 1910, Gary's African Americans, numbering fewer than 400, were mostly gainfully employed in the mill, on the railroads, or in private households, yet the Gary Post-Tribune ran headlines such as "City to Get Rid of Worthless Negroes."[20] When Blacks arrived in considerable numbers during World War I, they had little initial conflict with White immigrants. But by the 1920s a clear pattern of racism had set in.

As early as 1908 the school superintendent had transferred almost all Blacks to a segregated school.[21] In 1926, plans were announced for an all-Black high school, although an existing high school was located less than one-half mile from the proposed site.[22] In 1917, U.S. Steel decided to build housing for Blacks only, setting the stage for residential segregation, a process that was completed by 1940. Until the 1930s, Blacks were not admitted to the two city hospitals and African American physicians did not have staff privileges. Gary had two city golf courses, one with 18 holes for Whites, the other, with nine holes for Blacks, located in a drainage area prone to floods. African American citizens were not able to use the Lake Michigan beaches and most other public accommodations until the 1960s. In 1934, African Americans threatened to use political power to correct injustices. The White response was to re-place the ten wards elected by district with a six-ward format with all members initially elected at-large, thus reducing the number of Gary's African American city council members from three to only one.[23] These actions by public officials and leading private business encouraged working-class Whites, most of whom were new arrivals to the United States, to adopt strong anti-Black attitudes that persist well into the 1990s.[24]

World War II spurred Gary's steel industry and returned prosperity to the city. After World War II the eastern and southern European ethnic groups joined together and began to attain political power within the local Democratic Party. Until the late 1940s, Republicans allied with U.S. Steel and local business and institutional elites had controlled Gary's politics. But in 1951, when Gary was at its economic height, citizens elected Democrat Peter Mandich, son of a Serbian-born fireman and a former All-American football player at Tulane University, as mayor. Now, for the first time in Gary's history, the city's political and business communities would part company.

PLANNING AND THE
ETHNIC POLITICIANS, 1951-1967

Peter Mandich's electoral support came from White ethnics but also from Gary's African American community, which by 1950 was 30 percent of the city's total population. At the time of Mandich's election, Gary was poised for economic growth in a manner unprecedented in its brief history. World War II was over and the nation longed for consumer goods such as automobiles, stoves, refrigerators, washers, dryers, and small appliances, all of which were made out of steel. Steel production was also needed to rebuild Europe's infrastructure, since most of that conti-nent's mills were damaged or destroyed and would not be rebuilt for another decade. Industrial diversification was an issue discussed not only in Gary but also in other steel-producing centers such as Chicago, Birmingham (Alabama), and especially Pittsburgh.[25] It was time for government and business leaders in Gary to come together and plan for the day when other nations had steel-making capability. Pittsburgh had initiated this process in the late 1940s, with Mayor David Lawrence and leading businessman Richard King Mellon forming an active public-private partnership, one that continues to the mid-1990s.[26]

However, Gary's Republican business community chose not to work with Mandich, instead tying him with organized crime.[27] The Women's Citizen Committee, a Republican-dominated group, blamed Mandich for initiating a crime wave. Mandich labeled this group "a bunch of old biddies whose real dislike was immigrant participation in politics."[28] Mandich continued to press for immigrant rights. In 1952 he fired Republican Police Chief James Trager for being "indiscreet and heavy handed in harassing immigrants."[29]

The response of Gary's business and institutional elites, U.S. Steel, the Gary Chamber of Commerce, and the *Gary Post-Tribune* was a quiet, unpublicized disinvestment from Gary.[30] After 1951, there was no major private construction in downtown Gary, and private builders began to experience real difficulty in obtaining bank capital.[31] While the rate of new housing construction in central cities nationally was one percent between 1950 and 1960, Gary's rate fell to less than one-half of one percent by the early 1960s. By 1960, only 15 percent of Gary's residential units were less than a decade old while in the surrounding suburbs, the rate was 33 percent. In 1950, citywide retail sales reached an all-time high with 16,000 jobs in that sector, including 14,000 in the city's core area and 8,000 in the central business district. In 1955 Village Mall, the region's first auto-oriented shopping center, opened in suburban Ross township just across the street from the Gary city line. By 1960, the number of retail trade jobs in Gary dropped to 10,000 citywide and only 6,000 in the central business district. Sales to suburban customers fell drastically and 25 percent of Gary residents' purchases were made outside of the city.[32] While Gary's population reached an all-time high of 178,320 in 1960, 70 percent of the 45,000 person increase from 1950 consisted of newly arrived African Americans, mostly from the South. Between 1950 and 1960, over 50,000 Whites left Gary for the suburbs.[33]

Reelected in 1955, Mandich vacated the mayor's office in 1958 to become Lake County Sheriff. He was succeeded by George Chacharis, a Greek immigrant, who had played a major role in Mandich's successful mayoral campaigns. In 1959, Chacharis was elected mayor in his own right, winning 72 percent of the vote in the general election. He received 90 percent of the Black vote and appointed Gary's first African American cabinet member, City Attorney Henry O. Schell.[34] Realizing that Gary faced serious decline, in 1960 Chacharis initiated the city's first public-private partnership, the Committee of 100, which encompassed a broad spectrum of governmental, business, labor, and institutional leaders, including four African Americans and three women. Among its recommendations was the establishment of a planning department to prepare a comprehensive plan. The Committee of 100 provided the salary for a professional planner, a secretary, office, space, and supplies for an eighteen-month period. After that, an agreement called for the planning staff to move into City Hall and be placed on the payroll with full benefits.[35] The staff conducted initial studies and in 1962 engaged the services of Tech-Search, a Chicago-area planning firm, to prepare Gary's first comprehensive plan. Completed in 1964, this 140-page document contained a two-page policy statement and detailed plans for land use (residential, commercial, industrial), public facilities, transportation, and implementation. It was essentially a state-of-the-art physical development scheme modeled after the highly successful plan for Philadelphia prepared by Edmund Bacon and Larry Reich in 1961. However, it lacked two essential features of the Philadelphia plan: (1) an urban design component and (2) a detailed plan of action for revitalization of the central business district. The entire plan contained not one word about urban design, although Kevin Lynch's classic *Image of the City* was published in 1960 and was immediately read and reviewed by numerous architecture and urban planning publications. The central business district received one half-page with the following notation:

> The central business district has many varied and complex problems. A revitalization is necessary to make the area a lively, safe, and pleasant place to shop. It should be the focal point of the Calumet Region for shoppers' goods such as clothing, furniture, appliances, and other "big ticket" merchandise.[36]

Obviously, the planning consultants did not read Raymond Vernon's classic *The Changing Economic Function of the Central City*, published in 1959.[37] Vernon forecasted that downtowns would retain the office function but would, in all probability, cede retail trade, residential, industrial, and cultural functions to the outlying suburbs. Nor did these planners review the New Haven Rotival Plan prepared for its downtown in 1953[38] or the downtown plans for Providence,

Rhode Island, and Fresno, California, all of which emphasized office use for the central business districts of cities similar to Gary in population size and complexity.[39]

But the major flaw of the 1964 Gary Comprehensive Plan was that it said absolutely nothing about the pressing social issues of racial discrimination in housing, employment, education, and public services. In 1960, Gary's population was 40 percent African American and by the middle of that decade, the racial split was almost fifty-fifty. While three of Gary's nine city council members were African American, very little had changed from the post-World War II days in terms of racial equality. Housing was still rigidly segregated, with most African Americans crammed into the Central District and living in aged slum housing. Schools were segregated by race due mainly to housing patterns but also because strangely gerrymandered attendance districts ignored geography to encompass and surround areas that were either majority Black or majority White. Access to skilled and professional jobs in the mills and private business was barred to African Americans. City employment was not much better. African American votes had provided the margin of victory for both Chacharis in 1959 and his eventual successor, Martin Katz, in 1963, yet by 1967 Blacks, 50 percent of the population, held only 28 percent of city jobs and were only 5 percent of the police force.[40]

Issues of racial discrimination could have been addressed in the comprehensive plan but were not. Because of housing segregation, realtors were holding large tracts of vacant land in the city's west side off the market in anticipation of a White "back to the city movement" that never materialized. If the plan had adopted open occupancy as a policy, this would have led the way for the west side to be developed for whatever group wished to reside there, which did happen by the late 1970s when Gary was over 70 percent African American and open housing was the law of the land. However, the planners simply colored the vacant west side land yellow on the "plan for residence" map, ignoring reality. The "plan for schools" also side-stepped the issue of integration in their one-page statement on school objectives. The resultant plan for schools and parks was based on acceptable radii for school attendance areas such as a one-half mile walking distance for elementary students. There was no attempt to situate new or reconstructed schools in a manner that would promote integration.

One might take the view that in 1964, dealing with racial integration issues within the context of a physical comprehensive plan would have been political suicide. However, the plan's failure to deal with race became its undoing. Three White city council members equated planning with communism; the other three were more open to planning, realizing that the city's economic future depended on forward thinking. The three African American city council members, led by Richard Gordon Hatcher, president of the city council and the city's next mayor, were hostile to the plan, because it *did not* address racial issues. While the plan was adopted by the City Plan Commission, it floundered in city council, never received support from more than two White city council members, and was eventually tabled.[41] If the plan had received full support by the three African American council representatives it would most likely have passed, according to Cleo Wesson, who served as a city council member from 1954 to 1991.[42]

Urban renewal also ran into racial roadblocks. In 1961, the U.S. Housing and Home Finance Agency initiated the Community Renewal Program, which provided grants to cities to coordinate their various urban revitalization projects. Gary was successful in obtaining such a grant and by 1967 had produced a draft program for revitalization of Gary's inner-city residential neighborhoods.[43] The plan called for relocating up to 40,000 residents over a ten-year period. Most of those to be relocated were African Americans, Hispanics, or elderly White immigrants in their sixties and seventies. The planners glossed over the issue of relocation, failing to take into account the additional difficulties of resettling African Americans due to rigid racial discrimination in housing.[44] City Council President Richard G. Hatcher, other African American city council members, and community leaders repeatedly blasted this plan for its failure to tie renewal to open occupancy in housing.[45] With only lukewarm support from the city's business community, the renewal plan failed to pass in the city council; when the final report came out in January 1968, it was quietly placed on the shelf.

THE HATCHER YEARS: 1967-1971 AND AFTERWARD

In 1963, Gary's voters elected Richard Gordon Hatcher, a twenty-nine-year-old attorney, to an at-large seat on Gary's city council. Chosen council president immediately after being seated, Hatcher led the fight for open occupancy, the elimination of police brutality and an increase in the number of African Americans in city jobs. In 1965, the city council passed an omnibus civil rights ordinance that included open occupancy. Even though this ordinance contained virtually no powers of enforcement, its passage elevated Hatcher to the top of Gary's Black community and led Whites to brand him as a "Black militant."

In 1967, Hatcher decided to run for mayor. A three-way race included incumbent Martin Katz and White ethnic spoiler Bernard Konrady. Hatcher won the primary election by 3,000 votes out of a total of 51,000 ballots cast. Even though Hatcher won only 40 percent of the primary vote, he was initially considered a shoo-in to win the general election as Democrats outnumbered Republicans by a five to one margin. However, the Lake County Democratic Party organization, controlled by John Krupa of the old Mandich-Chacharis machine, refused to endorse Hatcher, calling him a "Black Power militant" and a "Kremlin agent." Instead, it covertly backed the Republican candidate, White businessman Joseph Radigan. Gary's business community and the *Post-Tribune* backed Radigan, but Hatcher narrowly won the general election in November 1967, by a vote of 40,000 to 38,000. While 96 percent of Blacks voted for Hatcher, only 10 percent of Whites chose to do so.[46]

Hatcher took office on January 1, 1968. He faced problems of poverty, substandard housing, crime, and racial polarization. He had very few tools to fight these problems. The 1964 Comprehensive Plan was essentially a collection of colored maps lacking both a revitalization strategy for the city as a whole, especially the central business district, and a social policy approach to racial segregation in housing, education and employment. The 1967 Community Renewal Program was "Negro Removal" at its worst as it called for the uprooting of 40,000 Garyites with virtually no provision for relocation housing. The business community had endorsed Hatcher's Republican opponent and for some fifteen years had turned its back on Gary's White ethnic mayors and city government. Gary's businessmen now adopted a "wait and see" attitude toward Hatcher. Gary's new mayor could not count on their support, unlike Carl Stokes, Cleveland's first Black mayor (elected in November 1967), who was initially embraced by his city's business elite.[47] Hatcher wanted to prepare a new comprehensive plan but decided not to spend the necessary time and resources because his constituents demanded immediate action on social problems.[48]

The administration of Lyndon B. Johnson stepped in with millions of dollars in federal aid for Hatcher and Stokes, Democrats, and the first African American mayors of major U.S. cities. Between 1968 and 1972 Hatcher received over $100 million in federal funds for urban renewal, the Model Cities program, and antipoverty efforts. By 1971, almost 3,000 units of slum housing were demolished, and with the 1968 Fair Housing Act in place, relocation housing was much more readily available than in past years. Reconstruction was slow, but by 1983, the Hatcher administration had built 1,300 units of public housing and 1,100 units of moderate income section 235 and 236 housing, and it had placed 500 new or rehabilitated housing units in the Section 8 program.[49] By the 1980s, the percentage of African Americans holding city jobs rose to 70 percent compared to only 28 percent in 1967; and the percentage of African Americans on the police force rose from 5 percent in 1967 to 45 percent, including the police chief.[50]

In the 1971 Democratic Party primary, Hatcher's opponent was Dr. Andrew Williams, an African American physician and former Lake County Coroner. Williams was supported by the Lake County Democratic party machine and Gary's business community. Hatcher won the primary by taking 60 percent of the vote and then won a second term in the November general election. The eventual response by Gary's business elites was to move the entire downtown to the suburb of Merrillville. In early 1972, plans were announced for two shopping centers totaling over one million square feet, along with accompanying strip malls, and by 1978, the three anchor department stores and many retail establishments in the downtown had either closed altogether or moved to the new malls and strip centers, taking their tax dollars with them. The city's banks, insurance companies, and hospitals also moved their headquarters to the suburbs. The gulf between Gary's business community and the city's political leadership, which began after Mandich's election in 1951, widened drastically during the Hatcher years.

Hatcher easily won reelection in 1975, 1979, and 1983, but by the mid-1980s, the city was literally falling down around him. Employment in the mill dropped from a high of 30,000 in 1979 to only 6,000 by 1983. The Reagan administration drastically reduced federal aid to Gary, forcing Hatcher to lay off over 300 of the city's 1,200 workers in 1983. A vast array

of public works—including a convention center and a sports complex—
and a new comprehensive plan passed unanimously by the city council
in 1986 failed to generate enthusiasm among the voters. Hatcher lost the
Democratic Party primary in 1987 to insurgent Thomas Barnes, a former
political ally. Barnes, however, lost voter support when he failed in his
attempt to land a multibillion dollar federally financed airport. The city
lost 35,000 residents between 1980 and 1990, including some 14,000
African Americans, most of whom moved to other Midwestern cities and
the Sunbelt. While Barnes was able to win reelection in 1991, by 1995 the
city was virtually bankrupt and Barnes chose not to run for a third term.

CONCLUSION

At the beginning of this chapter, we set out to discern (1) whether
being a planned city from the outset help Gary to grow and develop in
an orderly, rational manner; (2) the most crucial time period for planning
decision making; and (3) the impact of planning and race dynamics on
the city's development.

Gary started as a planned community but U.S. Steel planned only half
of a town, the mill and the first subdivision. The city's growth, especially
from 1910 to 1930, was determined by real estate speculators. The initial
plan was extremely flawed as the mill took up almost the entire lakefront,
placing the downtown area and even the first subdivision behind the
smoke-belching mills. U.S. Steel's failure to use either the garden city
techniques of Ebenezer Howard copied by Metzendorf or the neo-
baroque scheme of Daniel Burnham doomed Gary to design mediocrity.
Ironically, while U.S. Steel's engineers were designing Gary in 1903-1906,
the company was a major contributor and booster of Burnham's majestic
Plan for Chicago.[51]

A review of Gary's history from 1900 to 1970 shows that the most
crucial period for planning decision making after the initial city's design
was 1950-1960. By 1950, the city was reaching its growth peak and poised
for its most successful period of economic development. This was the
time to plan for industrial diversification with related ancillary benefits
in housing, recreation, education, and cultural activities. Required were
high-level public-private partnerships, such as those forged in the immediate postwar years in Pittsburgh, Cleveland, Philadelphia, and New

Haven, Connecticut. However, the opportunity was lost as the Gary business community refused to work closely with the White ethnic city government headed by Mayors Peter Mandich, George Chacharis, and Martin Katz. Comprehensive plans and urban renewal strategies needed in the 1950s were not undertaken and completed until the mid 1960s. By that time, the opportunity for economic growth was lost; Europe had rebuilt its steel mills, and nations such as Japan, South Korea, India, and Brazil were beginning to develop thriving steel industries of their own.

Planning and race dynamics played a major role in Gary's growth and eventually its physical and economic collapse in the 1980s and early 1990s. Since Gary was essentially a one-industry company town dominated by U.S. Steel, this corporation could have led the way toward social enlightenment. Instead, it encouraged racial segregation and discrimination in the early 1920s to split working-class White ethnics and African Americans into competing groups, with the latter being used as a reserve army in a dual labor market.[52]

By 1950, institutional racism on the part of Gary's business and institutional leaders had helped to create three separate population groups: a WASP upper middle class, mostly Republican and conservative; a loosely knit coalition of ethnic Poles, Serbs, Croats, Italians, Greeks, and several other nationality groups, mostly Democratic Party members; and African Americans, 30 percent of the city's population with a growing political and social consciousness. By the 1960s, Blacks led by Richard Gordon Hatcher demanded equality. They attacked both the 1964 Comprehensive Plan and the 1967 Community Renewal Program for their failure to address issues such as open housing, integration in public education, and access to skilled and professional jobs in city government, industry and business. These plans, late by at least ten years, failed because of lack of support from the Gary business community and African American political leaders.

Hatcher was never able to establish a meaningful public-private partnership with the business community, especially after the latter group literally moved the city's downtown to the suburbs in the late 1970s. Gary's attempts at comprehensive planning and coordinated urban revitalization in the 1980s were a case of "too little too late." The window of opportunity posed by the 1950s closed. From the 1960s on, automation, the impact of foreign steel imports, and reductions in federal aid accelerated Gary's decline.

Gary's initial plans were flawed and only half complete, chiseling a physical form that haunts the city to this very day. Plans that should have been formulated during the 1950s were not, due to WASP-ethnic White conflict. The 1964 Comprehensive Plan and the Urban Renewal Plan of 1967 drifted off into oblivion due to racial conflict and lack of a strong public-private partnership. By 1990, Gary had lost over a third of its 1960 population. Population in the two-county metropolitan area fell slightly from 635,000 in 1960 to only 604,000 in 1990. Employment in the metropolitan area's steel industry dropped from 70,000 in 1960 to 38,000 and the city approached bankruptcy. While planning in the 1950s with implementation by the early 1960s might not have eliminated decline altogether, the lack of planning, due mainly to ethnic and racial divisiveness, hastened Gary's fall and that of its region.

NOTES

1. See Gregory D. Squires, ed., *Unequal Partnerships: The Political Economy of Urban Redevelopment in Postwar America* (New Brunswick, NJ: Rutgers University Press, 1989); and Susan Fainstein, *Restructuring the City: The Political Economy of Urban Development* (New York: Longman, 1986).

2. Robert A. Catlin, *Racial Politics and Urban Planning: Gary, Indiana 1980-1989* (Lexington, KY: University Press of Kentucky, 1993), 30-42.

3. James Lane, *City of the Century* (Bloomington: Indiana University Press, 1979).

4. Statement by E. J. Buffington, president of the Illinois Steel Company, who was responsible for development of the new site, in *Gary Urban*, November 9, 1907.

5. *Gary Post-Tribune*, November 30, 1939.

6. Raymond A. Mohl and Neil Betten, "The Future of Industrial City Planning: Gary, Indiana 1906-1910," *Journal of the American Institute of Planners* 38 (1972): 203-13.

7. Arthur B. Gallion and Simon Eisner, *The Urban Pattern*, 6th ed. (New York: Van Nostrand, 1995).

8. Graham R. Taylor, *Satellite Cities: A Study of Industrial Suburbs* (New York: D. Appleton, 1915), 173-76.

9. Mohl and Betten, "Future."

10. Ibid.

11. Ronald D. Cohen and Raymond A. Mohl, *The Paradox of Progressive Education: The Gary Plan and Urban Schooling* (New York: Kenikat Press, 1979), 11-22, 35-66, 84-109.

12. Annual Reports, Gary Building Department, 1923-1930.

13. Annual Reports, Gary Engineering Department, 1924-1930.

14. Edward Greer, *Big Steel: Black Politics and Corporate Power in Gary, Indiana* (New York: Monthly Review Press, 1979), 82.

15. Ibid., 81.

16. *Gary Evening Post*, May 11, 1920 report from Bennett and Parson's Planning Consultants prepared April 1920.

17. *Gary Evening Post,* September 20, 1920.

18. James Quillen, *Industrial City: A History of Gary, Indiana to 1929* (New York: Garnett Press, 1986), 392-93.

19. Federal Writers Project, *The Calumet Region Historical Guide* (New York: AMS Press, 1938), 163.

20. Raymond A. Mohl and Neil Betten, "The Evolution of Racism in an Industrial City, 1906-1940: A Case Study of Gary, Indiana," *Journal of Negro History* 27 (1974): 54.

21. Cohen and Mohl, *Paradox of Progressive Education,* 110-14.

22. Ibid., 118-22.

23. Raymond A. Mohl and Neil Betten, *Steel City: Gary, Indiana 1906-1950* (New York: Holmes and Meir, 1986), 48-90.

24. Mohl and Betten, "Evolution," 54, 58-62.

25. Alberta Sbragia, "The Pittsburgh Model of Economic Development: Partnership, Responsiveness and Indifference," in *Unequal Partnerships: The Political Economy of Urban Redevelopment in Postwar America,* ed. Gregory D. Squires (New Brunswick, NJ: Rutgers University Press, 1989), 104-5.

26. Ibid., 107.

27. Lane, "City of the Century," 259.

28. Ibid., 251.

29. Ibid., 252.

30. H. B. Snider, publisher of the *Gary Post-Tribune,* letter to Robert Gasser, president, Gary National Bank, October 27, 1953, Gary Chamber of Commerce Papers, 1945-1967, Calumet Regional Archives.

31. Greer, "Big Steel," 97.

32. Ibid., 98.

33. U.S. Census, 1960.

34. James V. Lane and Ronald D. Cohen, *Gary: A Pictorial History* (Norfolk, VA: Doming Company, 1983), 174.

35. Catlin, "Racial Politics," 69.

36. "Comprehensive Plan: The Master Physical Development Plan for the City of Gary, Indiana," June 1964, 5.2.

37. Raymond Vernon, *The Changing Economic Function of the Central City* (New York: Committee for Economic Development, 1959), 40-62.

38. Norman I. and Susan S. Fainstein, "New Haven: The Limits of the Local State," in *Restructuring the City: The Political Economy of Urban Redevelopment* (New York: Longman, 1983), 37-40.

39. Not only are New Haven, Providence, and Fresno similar to Gary in size, they also have the same relationship in terms of time, distance, and hegemony to New York City, Boston, and San Francisco/Los Angeles respectively as Gary has to Chicago.

40. Greer, "Big Steel," 108-9.

41. Interview, former councilman Cleo Wesson (member of the Gary City Council from 1954 to 1991), August 12, 1986. This interview was conducted as part of the adoption process of Gary's 1986 Comprehensive Plan, prepared by Robert A. Catlin and the Gary Department of Planning and Development.

42. Ibid.

43. Gary Redevelopment Commission, "Community Renewal Program," by Community Planning Associates, initial draft May 1965, second draft, July 1967, final report, January 1968.

44. See Martin Anderson, *The Federal Bulldozer* (Cambridge, MA: MIT Press, 1964), 91-105; and Chester Hartman, "The Housing of Relocated Families," *Journal of the American Institute of Planners* 30 (1964): 266-67, 268-82.

45. William E. Nelson and Phillip Meranto, *Electing Black Mayors: Political Action in the Black Community* (Columbus: Ohio State University Press, 1977), 86-144.

46. For details of Hatcher's successful election campaign of 1967, see Nelson and Meranto, *Electing Black Mayors*, 270-312; Lane, *City of the Century*, 349-61; and Alex Poinsett, *Black Power: The Making of Richard Hatcher* (Chicago: Johnson Press, 1970), 80-89; Greer, "Big Steel," 90-98, 145-48.

47. Nelson and Meranto, *Electing Black Mayors*, 145-202.

48. Catlin, *Racial Politics*, 72.

49. Ibid., 198.

50. Ibid., 198-99.

51. Taylor, "Satellite City," 173-76; Mohl and Betten, "The Future of Industrial City Planning," 203-13.

52. Greer, "Big Steel," 72-19; Mohl and Betten "Evolution," 58-62; idem, "Steel City," 47, 50.

MODEL CITIES REVISITED

Issues of Race and Empowerment

JUNE MANNING THOMAS

The U.S. federal programs of importance to urban development fall into two broad categories: those designed to assist cities, such as urban renewal, Model Cities, or urban mass transportation grants, and those that are not necessarily "urban" but that have major effects on urban areas. Several well-funded "nonurban" initiatives helped finance the post-World War II suburban exodus from central cities, such as the Federal Housing Administration mortgage insurance program and federal highway programs.

Hundreds of thousands of people left central cities for suburbs that were closed to minorities yet subsidized with generous federal dollars. The supposedly helpful "urban" programs were not only poorly funded, they often appeared to be doing more harm than good. One example is urban renewal, in force from 1949 to 1974. With that program, cities bulldozed countless inner-city neighborhoods, many of them African American, with scarce attention to the social and economic needs of residents, with scant or nonexistent help for relocation, and with little chance for speedy return into project areas. Dubbed "Negro removal" by its critics, urban renewal benefited some (see Shipp, Chapter 11, this volume) but devastated many.

The 1960s brought urban programs that appeared more attuned to the needs of inner-city residents. In 1966, President Lyndon Johnson proposed creation of Model Cities, a program that would bring "cities of spacious beauty and lively promise, where men are truly free to determine how they will live."[1] Model Cities gave cities "block grants" to attack a variety of social, economic, and physical problems, under the guidance of the newly created U.S. Department of Housing and Urban Development (HUD). A corrective antidote to the oppressive side of urban renewal, the program was designed to appeal to the urban Black constituency, by encouraging residents to participate in defining and supervising neighborhood improvement and by funding multifaceted rejuvenation. At first, Model Cities appeared to be a marvelous innovation, a peace offering to the disaffected.

In reality, Model Cities suffered from inadequate resources, local political battles, uneven performance, and poor federal leadership. The program lost its presidential patron when Johnson left office. In 1974, Model Cities died and yielded remnants of itself to Community Development Block Grants (CDBG). Low-income central-city residents were not an important constituency for the Nixon administration; CDBG therefore became less participatory, more oriented toward physical development, and less targeted to distressed urban areas than Model Cities had been. Because it opened up eligibility beyond a few targeted cities, CDBG turned focus and financing away from poor Black urban areas and toward a wider range of more prosperous cities and suburbs (see Part 5.G, this volume). The CDBG program has survived for well over two decades, even as other programs have arisen and fallen, and finances city-sponsored housing redevelopmental and community improvements. Meanwhile, Model Cities, labeled a failed experiment—even though it offered major examples of Black community activism—has languished in the backwater of urban policy history.

It became important to reconsider the previously maligned Model Cities when Congress created Empowerment Zones/Enterprise Communities (EZ/EC) under Title XIII of the Omnibus Budget Reconciliation Act of 1993.[2] Although EZ/EC and Model Cities differed in many ways, the connections were unmistakable. Both required eligible communities to compete with each other for federal designation. Both encouraged local and federal agencies to collaborate and to target funds to winning communities. Both allowed a range of local programming options and focused on strong citizen participation.[3] If Model Cities had been so bad, why did EZ/EC mimic it so well?

Perhaps Model Cities really was not so bad after all. This chapter uses the alternative perspective of African Americans to show Model Cities in a new light. To structure this chapter's research I have used the framework of inferential qualitative analysis, in part to explore the potential of such an approach for historical studies of race and urban policy.[4] My examination of two case study cities suggests that Model Cities had several decided benefits, particularly for African Americans, and that its problems and benefits provide crucial guidance for today's programs.[5]

RETHINKING TRADITIONAL EVALUATIONS OF MODEL CITIES

Much of the record concerning Model Cities comes from several evaluation studies of the national program. These, however, had a number of flaws. HUD contracted with the consulting firm of Marshall Kaplan, Gans, and Khan (MKGK) to evaluate the program but did not allow long-term assessment of program effects. Most MKGK material focused only on the program's first two or three years.[6] So the first evaluation flaw was that the major evaluators could not finish studying the program.

The second flaw was methodological. MKGK collected an impressive volume and variety of information about case study communities. It did not use other, potentially more powerful research designs, however. For example, the official studies did not compare Model Cities communities with matched nonprogram communities or Model Cities target area populations with nontarget area populations.

The third area of concern is conceptual. The major evaluators looked at success as defined in set ways. Marshall Kaplan, one of the MKGK evaluators, collaborated with Bernard Frieden to write the most well-known overall assessment. They documented how congressional politics expanded a program intended to target generous resources to a few key cities into one assisting 150 cities, while using the same amount of money.[7] They discussed HUD's failed attempts to provide a strong planning and review framework or to offer sufficient technical assistance. They explained HUD's reluctance to carry out strong national evaluation studies, even though the agency required local studies. The two authors concluded that the effort was "halfhearted," and that Model Cities could not possibly have succeeded.[8]

All this was probably true, but the criteria by which they judged the program were clearly oriented to particular concerns. As the authors explained, HUD soon changed program goals from improving area residents' quality of life to building up local governments' capacities. Toward this end, HUD developed several performance criteria, which heavily influenced the MKGK studies and Frieden and Kaplan:

- *Innovation* in structures, processes, and planning
- *Mobilization and concentration of resources,* involving an increase in traditional patterns of resources aimed toward the targeted areas
- *Coordination* among local, state, and federal agencies regarding Model Cities
- *Institutional change,* particularly increased sensitivity and responsiveness to the Model Neighborhood Agency (MNA)
- *Citizen participation,* such as residents' membership on boards and committees, employment in the agency, involvement in planning, and so on

According to MKGK, program performance faltered for each of these criteria, although this varied by locality.[9] Millett, who examined citizen participation for MKGK first-round case studies, found great local variation in participation "success."[10] Washnis documented variation in local performance as measured by the factors just mentioned and by "product"—the level of service, efficiency, and quality of service provided. He concluded that some cities, such as Chicago, succeeded reasonably well.[11] Few other authors have praised Model Cities.[12]

Conceptual predispositions often shape researchers' approaches. It is possible to judge Model Cities using different criteria from those just listed. Consider the program not from the theoretical framework of federal policymakers but rather from the vantage point of area residents. They may have doubted that more government innovation, or coordination among agencies, would have improved their lives and their communities. They might have called for an alternative approach to evaluation.

REVISTING THE ORIGINS
AND CONTEXT OF MODEL CITIES

Race and poverty probably shaped Model Cities residents' priorities. Model Cities clearly targeted inner-city residents, many of whom were African American. The percentage of African Americans living in target areas of cities chosen for evaluations was high in Washnis's study, as well as in first- and second-round cities studied by MKGK.[13] Considering the high percentage of Black residents in Model Cities areas, what may have been their key concerns?

Before Model Cities, several federal policies wrought devastating effects on inner-city minorities, including the urban renewal program, highway construction, and public housing. The key designers of Model Cities referred directly to the flaws of previous urban programs and redevelopment efforts. They also acknowledged that federal programs had "accelerated the diffusion" of families and jobs and that housing policies had supported segregation and favored suburban growth over city revitalization.[14]

The Chicago Urban League's critique of the 1966 Chicago Comprehensive Plan (see planning-related documents, Part 5.D) highlighted these and other concerns. The report called for eliminating racial segregation, broadening access to residential suburbs, increasing low-income housing stock, reforming relocation, and improving ghetto schools. Authors praised advocacy planning and increased political power for Blacks as possible antidotes to traditional urban planning and suggested that young Blacks should become semiprofessional planners that the community could trust.[15] Occasional commentaries by additional Black authors offered other glimpses into Black perspectives. Lillard argued in 1969 that "ghetto" communities should plan for themselves, hiring and firing their staff, and building and maintaining private as well as public housing. In 1970, Graves suggested ensuring Black representation on planning staff and decision-making bodies and redistributing power to local citizens.[16]

These reports and commentaries suggest important criteria for judging Model Cities. An alternative evaluation framework is that Model Cities should have promoted upward mobility of target area residents and improvement of their communities. Perhaps the program should have achieved something like the following:

- *Citizen empowerment,* or more effective citizen influence, capacity, and self-determination than mere "citizen participation"
- *People-centered development,* or reform of the physical planning agenda, and ability to create coordinated social, economic, and physical approaches to urban reform
- *Affirmative action* in city agencies and politics
- *Effective program product,* as measured by Washnis, particularly as this concerned alleviating poverty and its effects among racial minorities
- Cumulative *improvement of inner-city communities,* perhaps the ultimate "outcome" measure, which meant simply that the programs should have led to better community life for target areas

This new list of evaluation criteria obviously could include many other variables, particularly the crucial issue of opening access to suburbs; this chapter, however, considers only those criteria listed earlier.

TWO CITIES

Three midwestern cities—Chicago, Detroit, and Cleveland—received both Model Cities and EZ funds (in Cleveland through special supplemental Empowerment Zone funds). These three therefore make good case studies for Model Cities and the application of that experience to EZ, at least in the Midwest. This chapter considers two of these cities, Detroit and Cleveland.

The two cities' programs were very different. Detroit's target area was fairly diverse, 52.7 percent African American, while Cleveland's area was 92 percent non-White.[17] Detroit's program operated throughout the six years of the national program; its efforts are reasonably well documented. Cleveland's program had a difficult time getting launched and did not receive funds until mid-1971.[18] Partially because of this, primary sources are more limited.[19] The contrast in program maturity and possible effectiveness offers an opportunity to compare a program that was reasonably successful getting started and operating with one that was much less so.

Citizen Empowerment

Examining citizen empowerment is relatively easy because "citizen participation" was one of the primary factors HUD-sponsored evaluations evaluated. One of the hallmarks of Model Cities was supposed to be its involvement of residents in local program planning and implementation. About the importance of such participation both federal officials and local residents agreed, for it lay at the heart of making local programs appropriate and effective. Yet local officials had a hard time implementing the concept of meaningful participation, much less empowerment. Millett found significant variation in participation, as measured by resident influence over hiring and firing, budget decisions, program operation, and program initiation. Smaller cities, with few ethnic minorities, had higher levels of participation.

Millett studied the first two years of several programs, including Detroit's, where he judged citizen participation disappointing. Mayor Jerome Cavanagh used no resident input in the application for first-year planning funds. After designation, residents responded angrily to their exclusion and to the lead role assigned to the city's urban renewal agency. This response surprised the mayor, who never again gave the program his full attention. Residents helped write a stronger citizen participation plan but established no decisive decision-making role in program planning.[20]

Other materials provide an explanation of what happened after the planning phase. According to one dissertation, Detroit experienced a multiyear struggle among three parties: the citizen membership of the unwieldy Detroit Model Neighborhood Citizens Governing Board (CGB), composed first of 108 elected people and later of those 108 plus another 108 appointees; the staff of the Model Neighborhood Agency (MNA), many of whom were middle-class bureaucrats/planners; and elected politicians in city hall, including three successive mayors and the Common Council. The clash between CGB and MNA at times paralyzed the program.

Even after that, class differences between MNA staff and CGB residents, lingering distrust of bureaucrats, scuffles over stipends for attending meetings, and acrimonious disagreements over program funding decisions, some of which came to blows, stymied the ideal of smooth citizen participation. Legendary battles—for example, over funding of a recreational program—wearied participants and ended up causing the MNA director to hire personal police protection. Successive CGB appeals to Common Council eventually wore down the council's willingness to intervene.[21]

Cleveland also turned out to be a battleground. HUD denied the Locher administration's application for first round Model Cities planning funds. After Mayor Carl Stokes took office in November 1967, he submitted, with token citizen input, a new application for second-round planning funds. When it won designation, Cleveland set up a board of target area representatives elected from twenty-nine districts, with "planning councils" in each district.[22] But the internal rivalry in Cleveland's Black community between the Hough and Central communities, which made up the target area, stymied cooperation. So did conflicts between vocal citizen leaders and city hall staff.[23] Mayor Stokes' lukewarm commitment did not help.

The resulting turmoil handicapped the program, which experienced several delays getting HUD to approve subsequent versions of its program plans. HUD granted several extensions, apparently "to maximize citizen involvement," but Cleveland did not submit its First Year Action Plan until February 1971, having stretched its one-year planning funds "to the limit."[24] Not able to spend even their first $9.3 million allocation, the city program had one elected citizen board dismissed, experienced several run-ins between residents and staff, and went through four directors. At one point a third mayor, Ralph Perk, announced that the ceaseless fighting would force him to take supervision of the program away from the citizen board and turn it over to target area ministers to run. Citizen leader Fannie Lewis pointed out that he had signed a contract to the contrary.[25]

Two lessons seem apparent from these two cities' experiences. One was that citizen participation was a difficult process. Part of the problem was managing meaningful cooperation, but internal conflict within inner-city communities was also a challenge. The second lesson is not to equate a bad "participation" experience with a bad "empowerment" experience. Even in the midst of all the turmoil, citizens in both cities did in fact gain power and control over Model Cities planning and implementation. This was a brutal rather than a pretty process, but this was perhaps inevitable given the acute unevenness of previous power relationships and the poor federal guidance about how to make collaboration less bloody. Ancillary power came to residents because of the boost Model Cities provided. For example, the most vocal citizen-activist in Cleveland's Hough area, Mrs. Fannie Lewis, won election to the city council in 1980.

That Model Cities participation was not so bad after all became evident when citizen activists realized how little citizen involvement was required for the new CDBG. In its reaction to CDBG in 1974, one civil rights group claimed that Detroit's CGB had gained considerable authority over the city's Model Cities operation, and the group openly mourned the death of citizen power that CDBG's weak participation structure heralded.[26]

People-Centered Development

People-centered development refers both to reform of the physical planning agenda and to coordination of various approaches to urban redevelopment. One measure of success could be whether a target municipality reformed its methods of physical redevelopment; another could be the creation of holistic, as opposed to unidimensional, program initiatives.[27]

Detroit's Model Cities program encouraged people-centered development, almost as a matter of definition under Model Cities legislation. But Detroit also shows that citizens used local Model Cities to make sure the urban renewal dragon was dead. The first major clash between Mayor Cavanagh and the citizens of Detroit's Model Cities program came over citizen participation, but the next was over whether residents could halt and review all urban renewal projects located in the Model Neighborhood area (which included almost the city's entire program).[28] Although they lost this fight, citizens thereby served notice of their presence and scrutiny. The CGB counteracted the traditional urban renewal bureaucracy by hiring people from the city's other agencies and by involving them in activities centered around social services.

One instructive encounter occurred in Detroit when the CGB challenged the planning commission's right to plan their Model Neighborhood. Backed by the power of Model Cities, the board wanted to hire its own staff to carry out all of its planning, overall as well as project-specific; it did not trust the city's planning staff because of the "sins of the past."[29] After several months, it became clear that the city's planners would carry out overall planning but that they would have to show greater sensitivity.

Even more important were the initiatives funded by Model Cities. Some of these combined social, economic, and physical planning by creating programs that treated whole persons rather than parts, in such areas as employment, education, health, social services, recreation, crime, transportation, and housing. Staff also redefined "social planning"; John Musial, a creative and aggressive sociologist, spearheaded studies that forced the local utility company to change its rate structure, promoted affirmative action in city government, and pushed for community governance, resident bids for city contracts, and social planning within the master plan.[30]

Cleveland's intent was to reform the physical planning agenda and develop more holistic strategies, but this was not as successful as Detroit's program. Program objectives targeted a mixture of areas, and programs announced in 1973 and 1974 included mental and physical health care, day care, home rehabilitation loans, and a "mini-bus" transportation program.[31] However, little evidence suggests that Model Cities directly affected the city's physical redevelopment agenda or actively involved city planning staff.

The lessons: Model Cities did encourage a mixture of physical and social programs, particularly in Detroit. Detroit's program also challenged traditional city development approaches and educated planners about new ways of planning. This may not have met the standard of people-centered development, but it indicated progress in that direction.

AFFIRMATIVE ACTION

The term *affirmative action*, which can refer to many things, here means greater access by ethnic minorities and target area residents to program decision-making bodies or to agency or municipal jobs. This is actually an extension of the "empowerment" process. According to this criterion, Detroit's program spurred important changes in Detroit. The multiracial Citizens Governing Board gave racial minorities and target area residents access to governance, although unfortunately these often fought among themselves. The agency hired a higher proportion of Black planners than was typical for the times, and several administrators were Black, including several directors.[32] MNA brought more Blacks and women into the Detroit Civil Service system, several of whom transferred into other departments at job levels that Blacks and females had not held previously. The social planning unit's strong affirmative action hiring plan for women and minorities became a model for other city departments, including the planning commission.

Detroit's program included a "Resident Employment and Training Unit," designed to hire Model Cities residents with program funds. In its first twenty months, this program hired 2,347 people. Unfortunately, according to MNA staff, poor administration and unclear standards opened the door for nepotism on the part of the CGB. In later years, as MNA took control of hiring, program management improved.[33]

Cleveland involved community leaders in Model Cities administration and included two women from the target area in writing the application. The elaborate citizen participation structure was also evidence of Black representation, as was the influence of several key community leaders. Newspapers announced 50 agency jobs in 1972 and 500 program jobs in 1973, with 800 others to be trained and 1000 referred, but the success of these projections is doubtful.[34]

The impact on Cleveland's bureaucracy is unclear. The opportunities may not have been as strong because Mayor Carl Stokes, an African American, took office in 1967. This was a visible symbol of victory that may have made Model Cities seem less important as a pathbreaker for minority groups. Stokes appointed a White student with little clout in city hall to head the Model Cities effort in his administration—an action that indicated the program's low priority in his eyes—and one of his four program directors, a Black man brought in from California, lasted only six months.[35]

In sum, Model Cities in these two cities improved access to policy influence for a wider array of people, particularly African Americans and residents of target areas. In Detroit, the program may have opened up new job opportunities and promoted affirmative action in city government. Job and bureaucratic impact were less clear in Cleveland, although the expanded influence of target area residents was notable.

Effective Program Product

Effective program product is difficult to assess twenty years after the fact. Very few contemporary national evaluations convincingly addressed this issue, although various national and local evaluations provided some insight. In 1975 Detroit staff wrote a overall program self-evaluation, as required by HUD. Since this was not an outside evaluation, it cannot be called objective; however, many of its conclusions rely on evaluation studies commissioned at the local level. Furthermore, the document appears to balance self-congratulation with self-criticism.

In general, some initiatives had been very successful, and others had failed. For each aspect of each program, staff offered analyses of what had gone right and what had gone wrong. Table 9.1 presents a summary of their comments, with a rough topology of their assessment of program performance for ten main components. According to their judgment, several program components offered important services. For example, 887 dwelling units received some assistance, educational programs helped 9,500 children, and a new health maintenance organization enrolled 8,500 people.

Counting numbers of people served or programs implemented is one of the poorest forms of evaluation, and these numbers do not indicate that local programs were effective. However, such figures do suggest that Detroit's Model Cities program distributed a number of services to area residents. These may not have been the best programs, or the most effective in resolving deep-seated urban problems, but from the perspective of residents they may have met perceived needs.

Little additional evidence is readily available concerning the effectiveness of Cleveland's program. In 1973, Cleveland assured HUD that the city had "salvaged" its First Action Year Plan, but meanwhile Cleveland had forfeited $42 million in Supplemental Funds.[36] The city experienced only one or two implementation years, with perhaps one or two years of transition to CDBG. This was a foreseeable disaster considering its four-year application and planning process. In May 1974, local newspapers reported that a "Cleveland Urban Observatory" had conducted an evaluation of the manpower program. That study called Cleveland's Model Cities a "Greek tragedy," marked by "lack of leadership, direction, training, planning and evaluation," and benefiting mainly consultants.[37]

Table 9.1 Detroit Model Cities Performance, 1969-1975 (Self-Assessment by Agency Staff)

Program Component	Mostly Successful	Moderate Success	Mostly Not Successful
Employment			
Placement			x
Training			x
Education			
Elementary school[a]	x		
Vocational	x		
School council support			x
Health			
Health maintenance[b]	x		
Drug addiction[c]	x		
Mental health			x
Social services			
Loans and grants		x	
Social planning	x		
Recreation and culture			
Recreation fund	x		
Summer programs	x		
Mobile pools	x		
Cultural arts			x
Crime			
Probation	x		
Legal services	x		
Intensive treatment	x		
Community officer	x		
Economic			
Chene-Ferry Mall	x		
Small business			x
Transport/communication			
Demand/response system[d]	x		
Lighting	x		
Newspaper		x	
Communication centers			x
Environment			
Planning		x	
Small parks		x	
Demolition		x	
Animal services		x	
Housing[e]			
Construction and repair		x	
Relocation		x	

SOURCE: Created by author from textual information available in Detroit Model Neighborhood Department, "Final Local Evaluation of Model Cities Program, Detroit, Michigan, 1969-75," March 1975, pp. 7-26.

a. The program assisted approximately 9,500 Model Neighborhood school children over five years.

b. Health maintenance organization enrolled 8,500 people, and reduced health care costs.

c. More than 1,500 individuals served since project began; 295 were detoxified.

d. Demand-response system served 7,500 riders per month in 1975.

e. The Detroit Model Cities program assisted 887 dwelling units, including 260 new units completed or under construction; 126 units that received FHA commitment for construction; 432 units awaiting FHA commitment; and 69 homes repaired. Other program components listed production units in terms of numbers trained, numbers hired, and so on.

Inner-City Improvement

The last alternative criterion was cumulative improvement in city target areas. Here one would expect problems, given the spotty record of implementation. As explained earlier, HUD abandoned the goal of improving the quality of life in the target area. Those who believed the original claim that the federal government would create a series of "model" cities would have been sorely disappointed by the results.

The question still remains, however, whether target areas experienced improvements during and after the program. Did target areas—perhaps not because of Model Cities but, rather, because of the cumulative effects of other federal and local efforts—witness improved quality of life?

Evidence is mixed. On the one hand, today's visitors to both areas would notice several recent physical improvements, particularly in Cleveland's Hough area;[38] in both cities new housing developments suggest progress. Both target areas have benefited from sustained efforts by nonprofit groups.[39] Concrete evidence of physical improvements is apparent in Figure 9.1, which compares census data for tracts in Model Cities target areas with tracts in the city as a whole. Both target areas experienced a precipitous drop in the number of substandard houses during the 1960s and 1970s. Federal programs that could have assisted with such a drop include urban renewal, Model Cities, and CDBG. Perhaps the "rolling parade" of government and nonprofit initiatives over a number of years was more important than any one program.

However, the drop may have also come from other forces. Between 1960 and 1990, Model Cities target area population dropped in Cleveland by 80 percent and in Detroit by 63 percent. During those thirty years, the number of housing units in Cleveland's target area fell 67 percent, by 19,000 units, while the number of Detroit target area units fell 45 percent, by 30,000 units. Both cities experienced reduced percentages of substandard housing, therefore, in the context of drastically fewer people and houses.

That both Detroit and Cleveland chose EZ target areas that overlapped at least part of older Model Cities boundaries suggests that these areas are still in distress. Figure 9.2 confirms that suspicion; the percentage of Model Cities target area families falling below the poverty level increased steeply in both cities, but particularly in both target areas, during the period from 1970 to 1990.

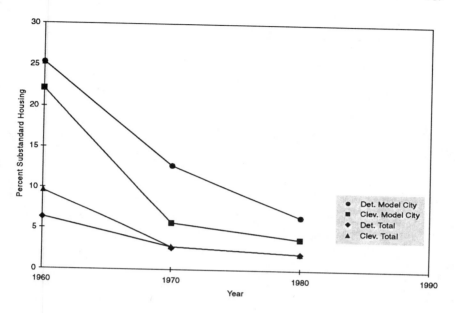

Figure 9.1. Percent Substandard Housing in Total City and in Model City Area, Detroit and Cleveland, 1960 to 1980

IMPLICATIONS FOR EMPOWERMENT ZONES AND OTHER PROGRAMS

Model Cities, obviously, was no panacea. It did not create the better cities that Johnson had promised. The two case study cities illustrate numerous problems and false starts. Yet a reassessment from a conceptual framework informed by the African American concerns suggests that Model Cities offered some benefits. It increased citizen empowerment, tamed the "dragon" of insensitive urban renewal, improved access to power, and sometimes implemented useful programs. As a national experiment in enlightened grantsmanship, the program faltered. This does not mean that every aspect of every local initiative failed or that the national program did not have valuable characteristics.

The experiences of Detroit and Cleveland do offer a number of lessons that are important to remember during contemporary attempts to create

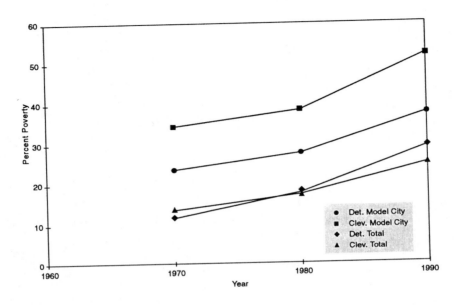

Figure 9.2. Changes in Family Poverty Level for Total City and Model City Area, Detroit and Cleveland, 1970 to 1990

effective urban programs. In particular, those involved in implementing or administering contemporary programs might consider the following points:

City Performance. One obvious lesson is that some cities may face extraordinary difficulties implementing new and innovative federal programs. The Cleveland Model Cities experience demonstrates that targeted communities, selected through national competition, can fail in their efforts to run new programs. Therefore cities in special urban programs such as EZ/EC should receive purposeful assistance from federal officials to help ensure that they succeed.[40]

People-Centered Programs. The Model Cities experience revealed that it was possible to extend beyond unidimensional approaches to urban reform and to combine social, physical, and economic programs in multifaceted efforts. But Model Cities was soon aborted, and techniques are rusty; CDBG facilitated some "holistic" solutions but did not allow as great a variety of programs. The hiatus in experience with such initiatives does not bode well. Localities and federal agencies will have to develop the skills necessary to implement the more multifaceted activities expected under programs such as EZ/EC.

Affirmative Action. Affirmative action may be less urgent in some cities in the 1990s than it was in the 1970s, especially considering the rise in the number of Black politicians and planning professionals. Yet it is important not to neglect this issue, since racial conflict is still very much alive, and access to power and authority for minority groups is still an important part of the empowerment process. Local and federal observers will need to be on their guard to ensure equal access by minority populations to local staff positions and governance structures.[41]

Beneficiaries. Programs must benefit their supposed constituencies. This may seem obvious, but in some cases Model Cities did this and was still judged a failure, since official evaluations undervalued service to area residents and overvalued government innovation and coordination. It is important to consider program effectiveness from the perspective of supposed constituents, as well as from the perspective of government. If contemporary programs do not meet the needs of the residents they are trying to serve, they are lacking, regardless of whether or not they improve performance of national or local governments.

EZ/EC attempts government innovation by requiring strategic and consolidated planning and by urging interagency cooperation at the federal level. Target area improvement, not government innovation, should be the primary goal of such a national program. EZ/EC also uses tax benefits to businesses and corporations as a major incentive for reinvestment and resident hiring. Such incentives must benefit minority-owned as well as other firms, and they should strive to bring lasting economic improvement (as opposed to temporary, short-term employment) in the lives of area residents.

Empowerment. Perhaps the most important lesson concerns citizen participation/empowerment. The noble goal of citizen participation proved very difficult to accomplish—handicapping Detroit's program and apparently killing Cleveland's—and the process of empowering citizens was fitful and painful. Many communities were surprised to find the participatory application and planning process for EZ/EC difficult and contentious. They should not have been surprised. In the years since Model Cities, urban programs required relatively weak citizen input, so cities had little experience in sharing power. The predictable outcome: EZ/EC brought a rude jolt, reminding us of the need to develop much more effective methods for citizen cooperation with local government.[42]

Evaluation. Finally, few exemplary national evaluations exist for Model Cities, even though HUD's administrative rules required evaluation at the local level. And so it is doubly important to undertake creditable national evaluation of current programs, to understand more clearly what works and what does not and why. Such evaluations must view the program from the perspective of the beneficiaries as well as the government.

NOTES

1. Bernard J. Frieden and Marshall Kaplan, *The Politics of Neglect: Urban Aid from Model Cities to Revenue Sharing* (Cambridge, MA: MIT Press, 1977), 51.

2. *Federal Register*, "Designation of Empowerment Zones and Enterprise Communities: Interim Rules and Notices," 59, 11 (January 18, 1994): pp. 2686-2712. Legislation allowed the secretary of HUD to designate six urban Empowerment Zones and sixty-five urban Enterprise Communities; the U.S. Department of Agriculture could designate three and thirty-five, respectively. A special category of designation brought limited benefits for Los Angeles and Cleveland. Empowerment Zones receive $100 million in federal Title XX funds over two years, as well as many tax benefits and special access to other federal program funds. Localities must carry out program activities in target areas in accordance with a guided ten-year strategic plan.

3. For strong comparisons of EZ/EC with Model Cities, see Elizabeth Gunn, "The Growth of Enterprise Zones: A Policy Transformation," *Policy Studies Journal* 21.3 (1993): 432-49; and Marilyn M. Rubin, "Can Reorchestration of Historical Themes Reinvent Government? A Case Study of the Empowerment Zones and Enterprise Communities Act of 1993," *Public Administration Review*, 54.2 (1994): 161-69.

4. See David Silverman, *Interpreting Qualitative Data* (London: Sage, 1993), who defines scientific inference for qualitative research and suggests how to pick a research design, what kinds of data to look for, and how to safeguard reliability and validity.

5. Special thanks to Vince Richardson, Michigan State University graduate student, for his research assistance.

6. MKGK wrote some studies credited to HUD, such as Office of Community Development, U.S. Department of Housing and Urban Development (HUD), *The Model Cities Program: A History and Analysis of the Planning Process in Three Cities* (Washington, DC: U.S. Government Printing Office, 1969); and HUD, *The Model Cities Program: A Comparative Analysis of the Planning Process in Eleven Cities* (Washington, DC: U.S. Government Printing Office, 1970). See Frieden and Kaplan, *Politics of Neglect*, 77n and chapter 8.

7. Frieden and Kaplan, *Politics of Neglect*, 37; Robert C. Wood, letter and report to Lyndon B. Johnson, November 30, 1964; Walter P. Reuther Collection, Walter P. Reuther Library of Labor and Urban Affairs, box 406, folder 4.

8. Frieden and Kaplan, *Politics of Neglect*, 233.

9. Office of Community Development, Evaluation Division, HUD, *The Model Cities Program: Ten Model Cities: A Comparative Analysis of Second Round Planning Years* (Washington, DC: U.S. Government Printing Office, 1973), 12, 84-91; HUD, *The Model Cities Program: Eleven Cities*, 70.

10. Ricardo A. Millett, *Examination of "Widespread Citizen Participation" in the Model Cities Program and the Demands of Ethnic Minorities for a Greater Decision Making Role in American Cities* (San Francisco: R & E Research Associates, 1977).

11. George J. Washnis, *Community Development Strategies: Case Studies of Major Model Cities* (New York: Praeger, 1974).

12. HUD, *The Model Cities Program: Ten Model Cities*, 31. HUD, *The Model Cities Program: Eleven Cities*, 23.

13. See Robert Woods's defense of Model Cities in "Model Cities: What Went Wrong—The Program or its Critics?" in *Neighbourhood Policy and Programmes: Past and Present*, ed. Naomi Carmon (New York: St. Martin's Press, 1990); and Rufus P. Browning, Dale Marshall, and David Tabb, *Protest Is Not Enough: The Struggle of Blacks and Hispanics for Equality in Urban Politics* (Berkeley, CA: University of California Press, 1984).

14. "Draft Report of the Task Force on Urban Problems," December 1965, 5, 6, Walter P. Reuther Collection, Walter P. Reuther Library of Urban and Labor Affairs, box 407, folder 5; see also Frieden and Kaplan, *Politics of Neglect*, 58, 59.

15. Harold M. Baron, ed., "The Racial Aspects of Urban Planning: Critique on the Comprehensive Plan of the City of Chicago," A Chicago Urban League Research Report (1968), 8, 12, 31; Walter Stafford and Joyce Ladner, "Comprehensive Planning and Racism," *Journal of the American Institute of Planners* 35 (1969): 68-79.

16. Leo Lillard Jr., "Model Cities, Model Airplanes, Model Trains," *Journal of the American Institute of Planners* 35 (1969): 102-4; Clifford Graves, "The Planning Agency and the Black Community," *Planning 1970* (Chicago: American Society of Planning Officials), 331-37.

17. City of Detroit, "Story of Model Neighborhood Program," brochure, n.d., and City of Cleveland, "Application to the Department of Housing and Urban Development for a Grant to Plan a Comprehensive Model Cities Program" (April 15, 1968), 2-10.

18. Cleveland, "Model Cities, Cleveland: First Year Demonstration Plan," (October 1972), frontispiece; Charles Morton, letter to "Residents' Board of Trustees and Executive Committee," September 16, 1970, in "Model Cities, Cleveland" (Cleveland, OH: Model Cities Association, n.d.).

19. A related commentary on Detroit's Model Cities is available in June Thomas, *Redevelopment and Race: Planning a Finer City in Postwar Detroit* (Baltimore: Johns Hopkins University Press, forthcoming, 1997). Sources on Model Cities in Detroit include a

dissertation, many city and consultant reports, newspapers, and correspondence/papers in local archives. Written sources on Model Cities in Cleveland include a few city documents and newspaper references (particularly *The Call & Post*, the African American newspaper).

20. Millett, *Examination of "Widespread Citizen Participation,"* 99-110. The MKGK study (HUD, *The Model Cities Program: Eleven Cities,* 70) also characterized citizen participation in Detroit as low during the two planning years.

21. David Cason Jr., "Social Class and Citizen Organization: Community Agency Interaction in Urban Planning Decision-Making" (Ph.D. diss., University of Michigan, 1976), 101-10.

22. HUD, *The Model Cities Program: Ten Model Cities,* 108-9. See also "City Seeks Model Cities," *Call & Post,* April 20, 1968, 1A; Alvin Ward, "Model City Candidates Get Constitution Briefing," *Call and Post,* December 14, 1968.

23. "Model Cities Program May Lose $9.3 Million," *Call and Post,* January 23, 1971, 8A.

24. Quotes Morton, letter to "Residents' Board of Trustees."

25. "Model Cities Election Still Being Counted," *Call and Post,* July 15, 1972, 7A; "Fannie Lewis is Named Citizens Participation Director," ibid., December 2, 1972, 18A; "Model Cities Will Determine," ibid., April 7, 1973, 1A; "Model Cities," ibid., April 14, 1973, 17A.

26. Michigan Advisory Committee to the U.S. Commission on Civil Rights, "Civil Rights and the Housing and Community Development Act of 1974," vol. 2: "A Comparison with Model Cities" (Lansing, MI: Advisory Committee to the U.S. Commission on Civil Rights, 1976), 29.

27. Washnis, *Community Development Strategies,* 14, discusses these kinds of issues in his review of "innovation," but also includes concepts such as creating programs that had not been tried before. HUD defined "innovation" as anything new to the municipality, particularly in its planning approaches. HUD, *The Model Cities Program: Ten Model Cities,* 12.

28. City of Detroit, Mayor's Committee for Community Renewal, "Proposal for Progress: Detroit's Application for a City Demonstration Planning Grant," March 1967, II-23.

29. Model Cities Housing Committee, memo to Manattee Smith and the Citizens Governing Board, August 19, 1969, Burton Historical Collection, Detroit Public Library, box 13, "Model Neighborhood Correspondence" folder.

30. Detroit Model Neighborhood Department, "Final Local Evaluation of Model Cities Program, Detroit, Michigan 1969-1975" (March 1975), 29, 36.

31. Program objectives listed in "Model Cities, Cleveland," 1970 draft. Program revisions listed in "Model Cities, Cleveland: First Year Demonstration Plan" (October 1972); "Model Cities Fills the Gap," *Call and Post,* March 10, 1973, 1A; "Model City Contract," ibid., August 25, 1973, 8A; "Model Cities Offers Loans," ibid., September 7, 1974, 3A; "Mini-Buses," ibid., September 14, 1974, 7A.

32. David Cason Jr., interview, June 18, 1989; HUD, *The Model Cities Program: Eleven Cities,* 41.

33. Detroit Model Neighborhood Department, "Final Local Evaluation," 34, 12, 29, 23.

34. "Model Cities Jobs," *Call and Post,* January 6, 1972, 4A; "Model Cities Leaders," ibid., February 24, 1972, 3A; "Manpower Center," ibid., March 3, 1973, 5A; "Model Cities," ibid., April 14, 1973, 17A; "Angry Afro Set Wrecks Offices," ibid., May 12, 1973, 1A, 4A.

35. HUD, *The Model Cities Program: Eleven Cities,* 108-10.

36. Robert Doggett, letter to HUD, Columbus office, June 1, 1973, cover letter for "Revision of the First Year Comprehensive City Demonstration Program," submitted to HUD, June 5, 1973.

37. "Model Cities Program Rapped," *Call and Post*, May 4, 1974, 1A, 4A.

38. Robert L. Joiner, "Urban Transformation," *Emerge*, September 1995, 32-39.

39. Particularly noteworthy in Cleveland have been the extensive, continued efforts of Fannie Lewis and the Hough Area Partners in Progress (HAPP), an active nonprofit community-based organization.

40. As of 1996, it appeared that the federal government was trying to provide such assistance to EZ/EC communities, through national conferences and frequent visits by federal officials.

41. One roadblock to smooth citizen participation in Detroit's recent EZ application process was the issue of equal access of all races to decision-making processes. Representatives of EZ area Black churches were particularly dissatisfied with the level of their involvement.

42. June Thomas, "Applying for Empowerment Zone Designation: A Tale of Woe and Triumph," *Economic Development Quarterly* 9 (August 1995): 212-24.

PART THREE

AFRICAN AMERICAN INITIATIVES AND RESPONSES

See Part 5.B, C, D, G, and J,
for planning documents
related to African American
initiatives and responses.

CHARLOTTA A. BASS, THE *CALIFORNIA EAGLE*, AND BLACK SETTLEMENT IN LOS ANGELES

JACQUELINE LEAVITT

harlotta A. Bass (née Spears) arrived in Los Angeles in 1910, and by 1912 she was managing editor and publisher of the *California Eagle*, Los Angeles's oldest Black newspaper. From 1912 until 1951, she used the newspaper as a platform for issues concerning race matters and documented the struggles around discrimination and segregation in employment and housing.[1] In 1952, she became the first Black woman to run for national office, as a candidate for vice president on the Progressive Party ticket. When she died in 1969, her leadership in struggles against restrictive covenants had helped change the pattern of Black settlement in Los Angeles. This chapter discusses Bass's personal and professional history as a means of understanding the fight against residential discrimination in Los Angeles. She deserves recognition. Along with many others outside the planning profession, she influenced the shape and quality of city life.

Los Angeles's treatment of race was not divorced from the national context. The combination of the significance of race in national politics, the importance of the Black press in reaching a broader public, and the extent of Bass's personal and professional networks, meant that local events inevitably reached a wider audience. One author observed that Bass's analysis of the interconnectedness of domestic and foreign affairs in the 1930s and 1940s was not made again until the civil rights and anti-Vietnam War movements of the 1960s. Today, few know of her significance in Los Angeles's history, and even less is known about her contributions on the national and even international scene.

Charlotta A. Bass was highly visible in her time. She was a public figure in a rapidly growing city, involved with a newspaper that, regardless of ranking and its ultimate demise in 1965, was a significant institution in Black life.[2] She was a national figure in the 1952 electoral campaign for the second highest office in the United States, despite the small number of votes her party received. Little has been written about her. This may be explained by her reticence, either as a personality trait or as expected behavior for women of her age, race, class, and status. Her memoir is more about the newspaper and race issues than autobiographical, and few other sources seem to exist. One likely reason that Bass's story is not better known is her politics. Bass's positions on domestic and foreign affairs were sympathetic with socialist and communist platforms of the day. The paper carried stories of the most turbulent events, whether labor sitdowns or strikes, riots or demonstrations.

Bass was exposed to the speeches and writings of the leading Black intellectuals of the day, including Marcus Garvey and W. E. B. DuBois. She attended the Pan-American Congress that DuBois convened in 1919 and participated in forming local civic groups around issues of race rights, such as the Los Angeles chapter of the National Association for the Advancement of Colored People (NAACP).

Bass arrived in Los Angeles the year before Job Harriman became the Socialist candidate for mayor of the city, a post he was expected to win (but did not). At one rally, having been asked to speak in his support, Bass was pushed off the platform by other Blacks who told her that the community was already committed to vote for Harriman's opponent, George Alexander. Not one to be bullied, she wrote an article on why the community should support Harriman.

Bass's unsuccessful attempts at electoral politics included campaigning in 1945 for the Seventh District City Council seat; she won enough votes to force a runoff. In the mid-1940s, the California version of the House Un-American Activities Committee investigated Bass and branded her, along with acquaintances and friends, a "commie." In 1948, after a lifetime as a registered Republican, disillusioned by segregation within the party, and angered by the lack of commitment to civil rights and peace on the part of both Democrats and Republicans, Bass joined the Progressive Party and helped found the Independent Progressive Party in California. She was active in the presidential campaign of Henry Wallace (Franklin Delano Roosevelt's third-term vice president) and Wallace's running mate, Senator Glenn H. Taylor of Ohio. Wallace's defeat did not deter Bass from continuing to speak on behalf of the Progressive Party platform against poverty, war, and slavery, and for welfare and civil rights.[3] In 1949, the Los Angeles Progressive Party successfully directed its main efforts to electing the first Mexican American, Edward Roybal, to the city council.

In 1950, Bass agreed to run for the House of Representatives in the predominantly Black Fourteenth Congressional District on the Progressive Party ticket. Her platform was world peace, worldwide neighborliness, jobs for all, civil liberties, and security. Bass lost to Samuel W. Yorty, who would later become mayor of Los Angeles (and during whose tenure the Watts Rebellion would occur). In 1952, the Progressive Party nominated Vincent W. Hallinan, a civil rights attorney, for president.[4] Bass became the vice-presidential candidate; Paul Robeson nominated her and W. E. B. DuBois delivered the seconding speech. The Progressive Party never thought it would win in 1948 or 1952, but the second campaign received an even smaller share of the vote than the first, dropping from 2.38 percent to .23 percent, which was 140,000 out of 61 million votes cast. The witchhunts that began in California now gathered more strength with McCarthyism. Bass received her share of threats and endured innuendos and attempts to discredit her. One sorority withdrew its honorary sisterhood because of her alleged identification as a communist. In other cases, former friends, like writer Arno Bontemps, no longer continued to share the same beliefs. The newspaper lost advertisers, and Bass sold the paper in 1951. Joseph Bass had died earlier in 1934. For all those years, Charlotta hoped that John Kinloch—a nephew who had migrated from the East, worked at the paper, and was also involved in fighting discrimination—would become the editor and publisher. These plans were aborted when Kinloch died in combat in World War II.

FROM SOUTH CAROLINA
TO LOS ANGELES

Charlotta A. Bass was born in either 1874, 1879, or 1880 to Hiram and Kate Spears in Sumter, South Carolina.[5] Of their eleven children, Charlotta was the sixth, one of four girls. When she was twenty, she moved to Providence, Rhode Island, to live with her oldest brother, Ellis, and took a job at the *Providence Watchman* soliciting subscriptions. After ten years, diagnosed as suffering from exhaustion, she left for Los Angeles for what was to be a two-year leave. After arriving in Los Angeles, Bass sought a job with the *Eagle*. John J. Neimore, the editor and publisher, hired her at a salary of $5.00 a week. He was ill and Bass became involved in every aspect of the paper. On his deathbed in 1912, Neimore, aware that his daughter Bessie was not interested in the paper, asked Charlotta to promise to "keep the *'Eagle'* flying."[6] The new editor published her first issue on March 15, 1912.[7]

Bass wrote that her first eight months as "editor, publisher, distributor, pressman and janitor of *'The Eagle'* stand out in my life as 'the period of hard times.' "[8] During late 1913, Charlotta hired Joseph Blackburn (J. B.) Bass, a founder of the *Topeka Plaindealer* who had also operated the *Plaindealer* in Helena, Montana. Soon after becoming editor, Charlotta and he married. Her official title was managing editor but Charlotta was the driving force, and from this vantage point, appears to have been the more militant.[9] The Basses clearly agreed that the newspaper had to "stay continuously on the firing line," on issues of racial equality.[10]

In 1959, Bass self-published her memoirs. Having sold the paper in the early 1950s, in 1960 she relocated to Elsinore, California, where the city's library became the venue for her activism. She urged Jewish and Negro parents to teach their children about their history and encouraged people to vote. She remained an activist until her death in 1969. Whatever the reasons that Bass's life has not penetrated public consciousness, the current upswing in right-wing politics, and the use of non-race-based but nonetheless discriminatory practices, provides an opportunity to reflect on this extraordinary woman and her responses to similar issues of an earlier period.

We can only speculate on the source of Charlotta's militancy. Little is known about her parents, Hiram and Kate, but living in Sumter, South Carolina, it is likely that Charlotta's parents were aware of if not active in the postslavery struggles and in local political organizing. The Union League may have been one subject that Sumter's Black community debated. Eric Foner writes that "in the end of 1867, it seemed, virtually every Black voter in the South was enrolled in the Union League or some equivalent local political organization."[11] Between 1868 and 1880, a growing Black press reported on similar efforts that were occurring in different parts of the country.

A typewritten account of Charlotta A. Bass's life, anonymously written and not dated, described her mother taking her to a meeting of the NAACP in 1902.[12] The young woman was enthusiastic about discovering people who were trying to create a better life for others. Somehow she had obtained a copy of Edward Bellamy's book, *Looking Backward*, first published in 1888, and the anonymous writer described Charlotta as strongly influenced by Bellamy's vision of a good society.[13] The Spears family tried to discourage her from socialism on the grounds that it was not Christian and would get her into trouble.

Charlotta's formal education record is unknown. Some references describe her as having taken college classes at Pembroke (now part of Brown University in Providence), Columbia University, and the University of California at Los Angeles. What is definitely known is that she came of age and matured into young adulthood at a time when Blacks suffered bitterly from the backlash to Reconstruction and enactment of new codes of segregation. Lynchings were commonplace.[14] We can surmise that she circulated among the growing class of professionals in the Black community who were questioning the next steps the race had to take in an increasingly hostile environment. Ideological and personal debates had sharpened around strategies that were identified with Booker T. Washington and accommodation, or W. E. B. DuBois and confrontation. Charlotta was in DuBois's camp. She was aligned with those who believed in political organizing, driven by the belief that they needed their own institutions. Los Angeles did not replicate the worst practices of the South, but the city and state harbored a goodly number of pro-Confederates and Ku Klux Klan (KKK) sympathizers.

Black voluntary organizations were essential for providing services, staving off efforts to reinstate or to institute Jim Crow laws, and supporting the struggles for equal rights and full citizenship. Disagreements within the Black community crystallized around methods, with recognition that militant positions invited death threats. At thirty-something years of age, Charlotta stepped into this milieu and rapidly assumed a public role.

THE LOS ANGELES CONTEXT

In 1910, the year that Charlotta arrived, Los Angeles was about to enter one of its major growth phases. The city's public relations machinery wooed visitors with visions of paradise in a year-round temperate climate, omitting or downplaying natural and manmade disasters. San Fernando Valley, an arid desert, was about to become valuable land, irrigated with water transported from the Owens Valley 233 miles away, at the expense of its ranchers. Their property was secretly bought, along with land in San Fernando Valley, and unsuspecting Los Angelinos voted approval for building the longest aqueduct in the world.[15] Transportation was also necessary to increase growth. By 1910, the city had secured federal funds to build an artificial deep-water harbor. Business interests oversaw the building of an extensive intra- and inter-urban rail system. Pacific Electric Red and Yellow Cars reached west to Santa Monica, Venice, and the Pacific Ocean; east to Alhambra and Monrovia; south to Long Beach and Newport; and north to Pasadena, ultimately covering more than 1,000 miles.

The first inhabitants in the Los Angeles area were the Gabrielinos, a branch of the Native American tribe the Shoshoneans. The Franciscan missionaries and Spanish military did not overcome the Native Americans until the last quarter of the eighteenth century as Spain colonized Alta (Upper) California, the territory north of Mexico. In 1781, the Spanish governor, Phelipe de Neve, laid out a plan for El Pueblo de la Reina de los Angeles, the Town of the Queen of the Angels, and recruited forty-four settlers, of whom the majority were either Indians, Negroes, or of mixed blood. By 1790, a city census indicated the shift in the population's race and ethnicity: the 141 people included 72 Spaniards and one European; the "minority" were 7 Indians, 22 mulattos, and 30 mestizos. In 1822, Mexico, with the help of settlers who had arrived from the United States, achieved independence from Spain. The territory once again changed hands when the United States fought and beat Mexico between 1846 and 1848. When the city of Los Angeles incorporated in 1850, only twelve Blacks were residents among a total population that had not yet reached 1,500. The most prominent Black woman was Biddy Mason, who arrived in California in 1851.[16] Mason, a practical nurse, became a real estate entrepreneur in 1866 when she invested her savings in two downtown lots between Spring and Broadway, Third and Fourth Streets.

By 1870, the Black population of ninety-three was a fraction of the total population of 5,728. In 1880, the city's population almost doubled to about 10,000, but the Black population barely grew to 102 and dropped as a percentage of the total population. In 1888, a Black community was identified for the first time at First and Los Angeles Streets. The number of Blacks climbed after that, although the percentage of the total population remained small: 1,258 Blacks were in Los Angeles in 1890, and the number doubled to 2,131 or approximately 2 percent of the 102,000 total in 1890. Two years later, the center for the "Negro" community moved east on Weller Street and south on First and Second Streets, with a business district developing at Second and San Pedro Streets. J. B. Loving, a leading Black realtor, wrote in the January-February 1904 issue of *The Liberator*:

> The Negroes of this city have prudently refused to segregate them-
> selves into any locality, but have scattered and purchased homes in
> sections occupied by wealthy, cultured white people, thus not only
> securing the best fire, water and police protection, but also the more
> important benefits that accrue from refined and cultured surround-
> ings.[17]

By 1904-1905, Black settlement moved south on San Pedro as well as east on Fifth and Sixth Streets. A year later, although Blacks were still able to locate anywhere in the city, the concentrated area was larger, bounded by Ninth Street on the south and Fourth Street on the north and from Central Avenue west to Maple Avenue. Up until 1910 or so, Blacks continued to be able to live anywhere in the city and Los Angeles boasted of its high Black home ownership rate (36.1 percent).[18] Marshall,

Texas, and Atchison, Kansas, were the only other cities to have greater homeownership among Blacks, at 37.5 percent and 53.6 percent respectively. The Black press publicized the widespread availability, affordability, and quality of the housing stock.[19] Lawrence B. de Graaf writes that the *California Eagle* was a notable exception.

> While Los Angeles had many appealing characteristics, the migration of Negroes to that city did not receive the same stimulus that Northern cities offered. Letters from friends and relatives encouraged many Blacks to come to Los Angeles, but its main Negro journal, the *California Eagle*, remained quite aloof from urging out-of-state readers to move to the city. The Negroes' condition in Los Angeles was criticized almost as much as theirs in the South, and a few articles discouraged movement to the city.[20]

Charlotta A. Bass was not alone in this position. W. E. B. DuBois came west in 1913 to encourage greater participation in the newly formed NAACP and to see for himself the opportunities in Los Angeles. He acknowledged the pluses but noted that "Los Angeles is not paradise."[21] Where others bragged to their Southern relatives that racism did not exist in California, DuBois pointed out the discrimination in department stores, hotels, and restaurants. These were all issues that Charlotta battled both through organizations she helped found and on the pages of the *California Eagle*.

CHARLOTTA BASS, THE CALIFORNIA EAGLE, AND BLACK SETTLEMENT: 1910 TO 1951

"By 1914, the *California Eagle* had won the reputation of being a people's paper, fighting on all fronts for the rights of the Negro people and other minorities to enjoy complete civil liberties."[22] Charlotta held memberships in other organizations, each overlapping in members and goals on behalf of the cause. She was the lady president of Division 156 of the Marcus Garvey United Negro Improvement Association (UNIA) and, despite ideological divisions, was active in the NAACP. She also organized, and was elected the first president of, the Industrial Business Council, a group formed to fight employment discrimination in the businesses located on the Central Avenue corridor. Her marriage to Joseph B. Bass did not change her priorities. "The new combination of editor and managing-editor realized that there was no time for romancing—with the ever-growing population which was now definitely heterogeneous and fraught with problems which the very early settlers did not encounter."[23]

Between 1910 and 1920, racial intolerance escalated and most Blacks, unless they were light enough and chose to pass as Whites, had to live in restricted areas. Some scholars attribute this to the jump in numbers.[24] Los Angeles's Black population increased from 7,599 in 1910 to 15,579 by 1920. In 1930, Los Angeles' overall population passed 1 million. The Black population rose to 38,894 and were living in five noncontiguous areas,[25] east, north, southwest, and southeast of downtown (see Figure 10.1). Central Avenue was the most prominent ghetto at the time, stretching for about thirty blocks and extending east to the railroad tracks. Four- and five-room California bungalows were going up rapidly, selling in the $900 to $2,000 range, with $100 to $200 down payments and $20 monthly payments. At these prices, they were in reach of even the laboring class of Blacks. However, any downturns in the economy threatened their tenuous stability. In an interesting twist on the impact of numbers, Charlotta A. Bass argued that it was the greater numbers of White workers emigrating from the South between 1910 and 1912 that gave rise to greater discriminatory practices against Blacks.

Figures vary but it has been estimated that during this period, Blacks were spatially restricted to anywhere from 5 to 20 percent of the total land area of the city. De Graaf found that by 1930, "70 percent of Los Angeles' Negroes were concentrated in one Assembly District in the Central Avenue area, and most of the remainder lived in adjacent sections."[26] Robert C. Weaver said that in 1947, although non-Whites were 6 percent of the population, "they had to crowd into only 4 per cent of the dwelling units."[27] Charlotta A. Bass described the consequences:

> Because the Black masses were forced into restricted areas, they became victims of the landlords and of the real estate brokers. What would purchase a cozy nook in Hollywood, would only buy a tumbledown shack on the Eastside. So now it became the custom in

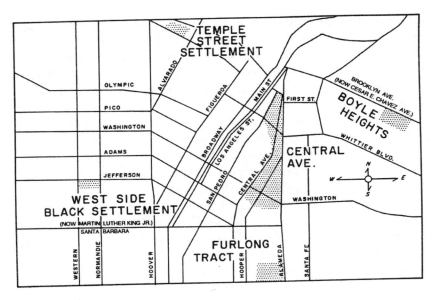

Figure 10.1. Black Communities in Los Angeles, 1920.
SOURCE: Emory J. Talbert, *The UNIA and Black Los Angeles: Ideology and community in the American Garvey Movement.* University of California at Los Angeles, 1980, p. 28.

Los Angeles for a great number of colored people to pay more for their inadequate, unsanitary, and ugly homes than the average White property owner paid for his moderately beautiful residence.[28]

In her memoirs, *Forty Years,* Bass recalled the early battles to fight such conditions. In 1914, Mrs. Mary Johnson bought and made the first down payment for a house on East 18th Street, off Central Avenue. At the time, East 18th Street was an all-White block. On the first day that Johnson moved in, having left the house for a few hours, "neighbors" removed all her belongings, putting them on the front lawn. They nailed the door tight and posted a sign: *Nigger if you value your hide don't let night catch you here.* Mrs. Johnson turned to the *California Eagle.* Charlotta A. Bass called some club women and "a brigade of a hundred women marched to the Johnson home" that night.[29] Unable to pry open the doors and windows that had been nailed "so securely that they couldn't make a dent in this 'fortification,' " they called the sheriff's office. The lawmen, who came only after a second call, removed the boards and let Mrs. Johnson reclaim her house.[30]

In 1925, the *California Eagle* published a letter written by G. W. Price, the Imperial Representative of the KKK. The letter outlined the ways in which Whites could gain control of Watts, at the time a separate city where Blacks had a good chance to win the next electoral campaign. Price sued for slander and the KKK tried to intimidate Bass, paying her a visit one night when she was alone in the office. They wanted an admission that the letter printed in the paper was fraudulent. She refused. The newspaper won the case but Charlotta later admitted that their next day's editorial heralding the end of the KKK was premature.

In 1929, the newspaper mounted an attack against Harry F. Burke, who was seeking a seat on the Los Angeles City Council. Burke was the head of the White Home Owners Protective Organization and his goal was to keep the area between Santa Barbara (now Martin Luther King Boulevard), Main, Manchester, and Vermont Streets as a White neighborhood. Whether the paper was the most decisive factor or not, its coverage was one element that figured in his defeat. A number of lawsuits resulted from Blacks deliberately buying houses in areas that included racially restrictive covenants. The *California Eagle* documented the attempts, the wins, and the losses, following the lawsuits as each case was fought separately.

LOS ANGELES AND COVENANTS

Individuals or groups used restrictive covenants in housing to bar other individuals or groups from residing in a house or area. Restrictions "might be placed in title deeds or signed as a separate agreement quite apart from the actual property titles," typically by land companies or persons with an ongoing interest in the real estate they sold.[31] A person or real estate firm could insert covenants into title deeds, but if a larger area was affected, another tactic was to get signatures from all the property owners mutually agreeing among themselves, and their heirs and successors, "to keep and observe particular restrictions against the sale or rent of their property to Negroes."[32] Charles Abrams identified twenty states, including the District of Columbia, which sanctioned racially restrictive covenants from 1918 to 1948, when the Supreme Court found them unenforceable. The litigation that wound slowly through state courts took different forms: White owners could sue White sellers, Negro purchasers, or both.

Many of the cases litigated in Los Angeles were part of a national strategy spearheaded by the NAACP that began in 1915 and lasted through 1953. Loren Miller, who became city editor for the *California Eagle* in 1932 (and would buy the paper in 1952), was a lawyer who argued many of the Los Angeles cases against racially restrictive covenants. Miller believed in a litigation strategy because "every time a Negro moved in . . . [t]he expenses of constant litigation are militating against those who desire to enforce the agreement."[33] At one point, twenty different suits were in progress in Los Angeles.

Looking back, one analyst questioned the importance of drumming up newspaper support to raise money or to stimulate "interest in the broader problems of Negroes in American society." He concluded that the publicity campaign was valuable because it served as an indicator that Blacks approached change through the legal system.[34] Charles Abrams, a planner and lawyer who fought for public housing and fair housing, believed that media coverage was a significant part of the strategy. Abrams criticized the mainstream media for not fighting discrimination and said they had mostly "proved indifferent to minority problems."[35] At the very least, the continuing coverage of the *California Eagle* must have bolstered the spirits and determination of those most affected by discriminatory practices.

In Los Angeles neighborhoods, particularly the ones near Black concentrations, "Keep Neighborhoods White" drives grew in popularity. In 1923-24, White homeowners tried to enforce a districtwide ban in the area east of Jefferson. This led to rising prices, and the area became available only to the wealthiest Blacks. In 1929, the California Supreme Court upheld the move by those fighting to "keep West Slauson White." Given such action by the California appellate courts, state decisions were no better than those at the municipal level and were "increasingly reactionary."[36]

Lawsuits continued their slow journey through the courts, and death threats continued when Blacks moved into houses in areas where Whites lived. Bass recalled that a renewed effort against segregation began around 1939. She attributed this to several causes, such as the presence in Los Angeles of John P. Davis. Davis spoke to Black leaders about steps necessary to ensure federal aid to Blacks.[37] His efforts were reinforced by news that real estate dealers in the San Gabriel Valley, east of downtown, had failed to arouse animosity by spreading rumors that soon Blacks "would own the entire community."[38] Another favorable sign was Superior Court Judge Georgia Bulloch's ruling, on November 15, 1939, that Mr. and Mrs. Sam Deedmon could legally live at 690 East 50th Street, "despite the fact that it was covered by racial restrictions, thirteen years old."[39]

In August 1941, White residents on Van Ness, Arlington, and Cimarron Streets, in the West Jefferson area of Los Angeles, spearheaded the fight against five Black property owners. Around the same time, KKK members targeted Blacks' homes as part of the "Keep West Slauson White" campaign. In the adjacent all-White city of Maywood, the KKK inspired a Race Restriction Program, publicized through an editorial on the front page of the Maywood-Bell *Southeast Herald*, and supported by clergy, politicians, and the majority of residents. The *California Eagle* printed a vehement response, citing the numbers of Blacks who had already died to defeat fascism, Marian Anderson's singing to raise sales of Defense Bonds, and China as an example of a country that was shaking millions free from slavery. The response ended with the words of a Southerner, Herbert Agar, editor of the *Louisville* (Kentucky) *Courier-Journal*, calling for the liberation of Black Americans.

The need for defense workers at home and abroad during World War II opened some doors for Blacks in employment and housing. But this did not occur without repeated fights. Wartime industries and federally subsidized "public war housing" finally eliminated discriminatory hiring and renting practices under threat of presidential executive order. Race baiting in this period was particularly vicious, fueled by the presidential executive order of 1942 that removed all Japanese and Japanese Americans from the West Coast. Little Tokyo became known as Bronzeville as Blacks moved into the relocatees' vacated apartments and houses.[40] In October 1943, following racial conflicts, the *California Eagle* formed a Negro Victory Committee, the purpose of which was to ask Mayor Fletcher Bowron "for a real city fight against discrimination."[41] Acting Mayor Orville Caldwell took advantage of the mayor's being out of town and issued an inflammatory announcement to "Stop the Negro Immigration." This pointed to "Little Tokyo" as a slum section appropriate to house newly arrived Blacks.

Bass described the interaction of lawsuits and the newspaper's involvement in battling discrimination in articles about the Laws family and "The House on 92nd Street." The Laws case provided an example of the problems arising when unsuspecting purchasers bought in areas that were restricted against Blacks. Both Laws and Lee Lofton had bought homes on East 92nd Street in 1923 and did not move to the sites until 1939. In 1942, prompted by realtors' discovery of the covenants, Judge Roy V. Rhodes issued a restraining order against the two families. This forbade them from occupying their homes, which they had lived in for three years. The Loftons chose to move from the neighborhood. The NAACP, which had taken the cases and which had refiled in a losing bid, advised the Laws family to leave in order to avoid facing contempt charges. The family decided to stay. They couldn't find another place; the weather was changing, and living in their automobile was impossible. Mrs. Texanna (Anna) Laws was not well but was afraid to disobey the court order. An anxious Henry Laws phoned Charlotta one day in 1944. She told him, "We'll fight this damnable condition until hell freezes over."[42] Charlotta A. Bass backed up her words by organizing a campaign and chairing the militant and integrated Home Protective Association. The newspaper rallied others who came out in support of the family. On November 30, 1945, Judge Allen W. Ashburn issued an ultimatum that the Laws either vacate the house in ten days or go to jail. A thousand people formed a protest line outside the house on the Sunday before they were escorted to jail.[43] At this point, national events overtook the Laws case.

California was not the only place where Blacks faced invasions of property rights, discrimination, harassment, and threats from the KKK and the White Citizens Council. At the same time that the Laws were fighting for their right to occupy the house they had bought, the U.S. Supreme Court agreed to review similar cases that originated in Missouri, Michigan, and the District of Columbia. The cases were known as *Shelley et ux. v. Kraemer et ux.* and *McGhee et ux. v. Sipes et al.*[44] Thurgood Marshall and Loren Miller were senior counsel for the petitioners. Charles Abrams wrote that the Court might have still avoided the issue but for the U.S. attorney general's highlighting President Truman's public statement that the federal government must be a "friendly vigilant defender of the rights and equalities of all Americans.[45] The timing was propitious, occurring not too long after the discovery of the horrors of German oppression and coinciding with publication of *To Secure These Rights*, a report by the President's Committee on Civil Rights.[46] Land use was not at the heart of the constitutional question; rather, restrictions concerned a designated class of people who were being denied access to own or make use of properties for residential purposes. The Shelleys were Black and, like the Laws, had purchased property without prior knowledge of restrictive covenants. The focus was on where state enforcement of private racially restrictive covenants violated the Fourteenth Amendment, which conferred rights to acquire, enjoy, own, and dispose of property. On May 3, 1948, the Court found that the racially restrictive covenants were valid between private parties but that the state had no enforcement powers. Chief Justice Vinson wrote the opinion for the Court:

> These are not cases, as has been suggested, in which the States have merely abstained from action, leaving private individuals free to impose such discrimination as they see fit. Rather, these are cases in which the States have made available to such individuals the full coercive power of government to deny to petitioners, on the grounds of race or color, the enjoyment of property rights in premises which petitioners are willing and financially able to acquire and which the grantors are willing to sell. The difference between judicial enforcement and nonenforcement of the restrictive covenants is the difference to petitioners between being denied rights of property available to other members of the community and being accorded full enjoyment of those rights on an equal footing.[47]

Before the cheering had stopped, it was clear that racially based struggles were taking other forms. In Missouri and Oklahoma, for example, the courts upheld suits for damages against the signer of the covenants. Bass used the *California Eagle* to alert readers to "property rights' abuses." Developers in newly opened parts of the San Gabriel and San Fernando Valley simply refused to sell to Blacks. Loren Miller predicted further troubles and noted that "there were many real-estate brokers and 'a substantial number of property owners organizations' in Los Angeles 'casting about for some way to enforce the covenants.' "[48]

In 1953, the Supreme Court agreed to hear the case of a White homeowner from Los Angeles who was suing for damages brought by other White property owners for violating a restrictive covenant. The NAACP lawyers who assisted Mrs. Jackson in *Barrows v. Jackson*[49] succeeded in arguing that she was able "to raise the issue that the rights of Negroes were being violated in the process."[50] Other than Chief Justice Fred M. Vinson, the Court interpreted the case as a means to close the gaps against racial covenants through finding covenants "unworthy" and barring damages. This was still insufficient to end discrimination. Since covenants were still allowed as private agreements, willing neighbors could ban Black entry by simply refusing to sell to a Black buyer as well as making it clear to the buyer that Blacks were unwelcome in the neighborhood. Most people would not want to move to a neighborhood under such circumstances, and thus the vicious circle of segregation was easily continued in many communities. In retrospect, as Charlotta A. Bass commented, restrictive covenants were only part of the problem:

> The problem of restrictive covenants thus became enmeshed in a multitude of other problems—housing appropriation bills in Congress, city planning work, problems of relocation, private financing of tracts, city ordinances and so on. The problem really boiled down to a broad fight for equality in many fields—to a fight for the acceptance of the principle of equality—social, economic and political—as something basic to the entire future of the land.

CONCLUSION

Charlotta A. Bass lived to see Blacks moving throughout areas of Los Angeles that had otherwise been forbidden through racially restrictive covenants. She witnessed integration and then resegregation as Whites fled, particularly from the South Los Angeles area. Conceding the prematurity of interpreting the Court's actions, Charles Abrams reflected on the effects of the covenant decisions.

> On the practical side the decisions released to Negroes some property which had been denied to them. These were not large tracts, nor were many in the suburban areas. But the Court rulings helped to break down resistance among owners of land bordering on minority-inhabited areas and speeded expansion into the adjoining sections and into some new sections. Much of this property would probably have

fallen into minority hands anyway. The decisions also removed the cloud from the title of properties already owned or occupied by Negroes who had bought or rented despite the covenants.[51]

Abrams added that, even if temporary, the moral climate was at least improved and may have contributed to the Court's favorable rulings against segregated transportation and educational facilities. On the debit side, segregation and discrimination did not fade away and housing conditions did not dramatically improve. The Federal Housing Administration (FHA), now part of the U.S. Department of Housing and Urban Development (HUD), under pressure from civil rights organizations and the attorney general, agreed to deny insurance to mortgages with covenants placed after February 1950. Of course, numerous other means to discriminate existed and segregation retained its tenacious hold.[52]

Bass died before the traditional Black communities in Los Angeles became predominantly Latino. The 1990 U.S. Census identified Los Angeles as being 13 percent Black and 39 percent Latino; currently the numbers of Latinos and Asian Americans are increasing. Within South Los Angeles, enclaves of Black homeowners still exist, and Blacks are a majority within the six public housing developments (of twenty-one citywide). Blacks who could afford to move left South Los Angeles after the Supreme Court's decision against racial covenants opened housing in the city's west side. By 1990, the so-called Black belt resembled a half moon that spread westward from South Los Angeles to Crenshaw and was thickest around Leimert Park and Baldwin Hills. Other Black settlements lie outside the city limits, within Los Angeles, Orange, San Bernardino, and Riverside counties. This dispersement could not have occurred without Bass's contributions to the fight against discrimination.

While racially restrictive covenants are now unenforceable, cutbacks in public services, abandonment of principles concerning the general welfare, attacks on publicly supported housing, and legally restrictive zoning ordinances limit people's mobility. Planning, a reflection of societal trends, is too often a tool to rationalize the new restrictive principles. In Charlotta's vision, however, planning as envisioned through progressive politics could make a difference: "Viewed in the long perspectives of history, our work [of the Progressive Party] will probably be regarded as a liberalizing influence on American thought."[53]

NOTES

1. Charlotta's tenure with the newspaper began prior to 1912, when she was John J. Neimore's assistant; hence, references to forty years.

2. Marianna W. Davis, ed., *Contributions of Black Women to America*, 2 vols. (Columbia, SC: Kenday Press, 1982); James Phillip Jeter, "Rough Flying: The *California Eagle* (1879-1965)," presentation to 12th Annual Conference of the American Journalism Historians Association, Salt Lake City, Utah, 1993.

3. Charlotta A. Bass, *Forty Years: Memoirs from the Pages of a Newspaper* (Los Angeles: Charlotta A. Bass, 1960), 145-53.

4. In 1995, the City of San Francisco, California, elected Vincent's son, Terrance, to district attorney and, at the same time, elected its first Black mayor, Willie Brown.

5. Sarah Cooper and Mary Tyler, Volume 1, *Encyclopedia of the American West* (New York: Macmillan, 1996); Kristine A. Yohe, in *Notable Black Women*, ed. Jessie Carney Smith (Detroit: Gale Research, 1992), 61-62, reports 1880. Bass's papers at the Southern California Library for Social Studies contain one reference to 1874 as her birthdate. Different locations reported for her birthplace include Sumter and Little Compton, Rhode Island.

6. Charlotta A. Bass, "A Brief Resume of the California Eagle," Christmas edition, 1927, unpaged.

7. Bass, *Forty Years*, 29-30.

8. Bass, *Forty Years*.

9. Jeter described Joseph B. Bass's arrival as essential to the continuation of the newspaper.

10. Bass, *Forty Years*, 33.

11. Eric Foner, *Reconstruction: America's Unfinished Revolution 1863-1877* (New York: Harper and Row, paperback edition, Perennial Library, 1989, 1st ed. 1988), 283.

12. The NAACP did not come into existence until 1910, and these events may have happened at other times.

13. Edward Bellamy's book received widespread popularity and influenced national figures such as Thorsten Veblen and John Dewey. Bellamy's critique of industrial capitalism and his proposals for forming cooperatives, eliminating poverty and unemployment, and creating equality in wage salaries, promised an alternative to economic depressions and labor unrest.

14. Herbert Aptheker, *A Documentary History of the Negro People in the United States*, vol. 2 (New York: Citadel Press, 1990), 651.

15. William Mulholland, chief of the city's Water Department, oversaw construction. Los Angeles annexed the 266-square-mile valley in 1915, doubling the city's land mass.

16. Mason secured her freedom through the courts in 1856, thus preventing her "master" from returning her to a slave state. Bass identified Mrs. Owens Bynum as the next Black woman to become a large-scale real estate entrepreneur, buying a tract in Boyle Heights that she sold and rented to Black residents.

17. Bass, *Forty Years*, 14.

18. Lonnie Bunch, "A Past Not Necessarily Prologue: The Afro-American in Los Angeles," in *20th Century Los Angeles: Power, Promotion, and Social Conflict,* ed. Norman M. Klein and Martin J. Schiesel (Claremont, CA: Regina Books, 1990) 101-30. On page 104, Bunch identifies the following neighborhoods: (1) West Temple Street to Occidental Boulevard; (2) First to Third Streets, San Pedro Street to Santa Fe; (3) Seventh to Ninth Streets, Mateo to Santa Fe; (4) Boyle Heights (First to Broadway, Evergreen to Savannah); (5) Thirty-Fifth Street and Normandie Avenue; and (6) Pico Heights.

19. Bunch, "A Past Not Necessarily Prologue," 101-30.

20. Lawrence B. de Graaf, "Negro Migration to Los Angeles, 1930-1950" (Ph.D. diss., University of California at Los Angeles, 1962), 21.

21. Bunch, "A Past Not Necessarily Prologue," 105.

22. Bass, *Forty Years,* 33. Other organizations included: the Los Angeles Forum, a club of influential Black men that began in 1902, and the Sojourner Truth Industrial Club that started in 1904 to "protect the welfare of Negro women by establishing a Christian, yet nonsectarian, dwelling that would be a safe refuge." In 1913, women were involved in forming the Los Angeles chapter of the NAACP.

23. Emory J. Tolbert, *The UNIA and Black Los Angeles* (Los Angeles: Center for Afro-American Studies, 1980), 87-108.

24. Clement E. Vose, *Caucasians Only: The Supreme Court, the NAACP, and the Restrictive Covenant Cases* (Berkeley, CA: University of California Press, 1959), 153; Charles Abrams, *Forbidden Neighbors,* 2nd ed. (Port Washington, NY: Kennikat Press, 1971), 217.

25. The areas were: (1) Boyle Heights (between Brooklyn Avenue on the north, Evergreen Cemetery on the east, Michigan Avenue on the south, and Mott Street on the west); (2) West Side Los Angeles, Jefferson between Normandie and Western Avenues, south to Thirty-Fifth Place; (3) Temple Street area, Beverly Boulevard on the south, Rampart Street on the east, Hyans Street on the north, Reno on the west; (4) Furlong Tract in the southeast section of the city; and (5) segments along and around Central Avenue.

26. de Graaf, "Negro Migration to Los Angeles, 1930 to 1950," 21.

27. Robert C. Weaver, *The Negro Ghetto* (New York: Harcourt, Brace, 1948), 123.

28. Bass, *Forty Years,* 96-97.

29. Ibid.

30. Ibid., 95.

31. Vose, *Caucasians Only,* 7.

32. Ibid., 8.

33. Ibid., 59.

34. Ibid., 73.

35. Abrams, *Forbidden Neighbors,* 333.

36. As stated by Loren Miller; see Vose, *Caucasians Only,* 62.

37. Bass, *Forty Years,* 101. Davis, with A. Philip Randolph, was active in the National Negro Congress, a group calling for racial solidarity, unionizing African American women workers, desegregating public accommodations and schools, protecting migrant workers, antilynching campaigns, and fighting fascism.

38. Ibid., 102.

39. Ibid.

40. Charlotta criticized the major Negro organizations for failing to establish fully cooperative working relations with the main organizations of the Japanese community; she hoped this would be overcome when younger Japanese leaders were allowed to return to the city.

41. Bass, *Forty Years,* 72.

42. Ibid., 109.

43. Ibid., 110.

44. *Shelley v. Kraemer*, 334 U.S. 1, 68 S. Ct. 836 was argued on January 15-16, 1948, and decided on May 3, 1948. *Hurd v. Hodge*, 334 U.S. 24 (1948), originating in the District, used a "fragment" of the Civil Rights Act in place of the Fourteenth Amendment.

45. Abrams, *Forbidden Neighbors*, 220.

46. Washington, DC: U.S. Government Printing Office, 1947, 31n in Abrams, *Forbidden Neighbors*.

47. *Shelley v. Kraemer*, in Derrick A. Bell Jr., *Civil Rights: Leading Cases* (Boston: Little, Brown, 1980), 85.

48. Vose, *Caucasians Only*, 240.

49. *Barrows v. Jackson*, 346 U.S. 249 (1953), in which the Court barred any award of damage.

50. Abrams, *Forbidden Neighbors*, 222.

51. Ibid., 223.

52. Ibid., 224-26.

53. Bass, *Forty Years*, 153.

CHAPTER 11

WINNING SOME BATTLES
BUT LOSING THE WAR?

*Blacks and Urban Renewal
in Greensboro, NC, 1953-1965*

SIGMUND C. SHIPP

The movement for slum clearance and downtown revitalization in U.S. cities was one of the elements included in Franklin D. Roosevelt's New Deal agenda, and it came to fruition with the passage of the Housing Act of 1937, which provided federal funds for the construction of housing projects in various parts of the country. The Housing Act of 1954, which specifically had "urban renewal" as an objective, called on cities to create comprehensive plans for rehabilitation, conservation, and demolition, not merely for a neighborhood as in previous legislation but for the entire city. Many liberal planners and legislators favored this approach to urban redevelopment because they questioned the ability and willingness of private developers to provide low-income, high-quality housing without some government assistance. In contrast, conservative politicians and business people disliked the idea of government intervention into the affairs of the private market, claiming that such involvement was socialistic.[1] Most city officials around the country welcomed the federal intervention and immediately set up groups and commissions to develop plans and programs suitable for federal funding. Since the focus was to be on slum clearance and redevelopment, Black residents were disproportionately affected by urban renewal programs because they were more likely to be housed in areas targeted for redevelopment. Researchers who have examined the impact of urban renewal on the Black population have focused on displacement and suggested that the entire program could be referred to as "Negro removal" or "Negro clearance." In his book *The Federal Bulldozer*, Martin Anderson estimated that between 1957 and 1961, almost two thirds of those persons displaced by urban renewal projects nationwide were Black or Puerto Rican. "Because of this," Anderson noted, "the program is sometimes referred to as a 'Negro clearance' program, with its goal being the creation or preservation of a White, middle-class neighborhood."[2] Researchers who examined urban renewal in Cincinnati, Buffalo, Pittsburgh, Philadelphia, and Boston came to similar conclusions.[3] Even those who favored the federal government's intervention in urban redevelopment, such as Jane Jacobs, author of *The Death and Life of Great American Cities*, came to criticize the impact of slum clearance on neighborhoods and their residents.[4]

More recently, researchers have begun to extend beyond "Negro removal" to examine other effects that urban redevelopment programs had on resident Black populations. Were there any elements within these urban Black communities that supported the coming of the "federal bulldozers"? Were there African Americans who participated in urban renewal decision making? Did some African Americans benefit from urban redevelopment programs, or were wealthy White developers and middle-class Whites the only winners? The later research indicates that certain segments of the Black population did benefit from urban redevelopment projects. For example, by the 1950s and early 1960s, African Americans in New York, Chicago, and several other Northern and Midwestern cities had acquired some degree of political clout within the Democratic Party and city administrations and were able to participate in policy-making regarding urban renewal. In most instances, however, Black Democratic appointees to redevelopment commissions endorsed the proposals put forward by their White colleagues.[5] Some middle-class

African Americans, as well as representatives of the local branches of the NAACP and National Urban League, spoke in favor of urban renewal programs because they saw them as ways to improve deteriorating Black neighborhoods and revitalize their communities.[6] Black protest did erupt in some places, but it had little impact on decisions to demolish buildings and homes in redevelopment areas. Jon Teaford recently re-examined urban revitalization programs in twelve cities in the Northeast and Midwest from the late 1940s to the early 1960s. While he found some Black input in decision making and even some protest, in general African Americans had little or no effect on the implementation of redevelopment projects.[7]   *[handwritten: But why? B/c of lack of political clout or lack of mobilizat or barrier to mobilizat]*

Comparatively less research exists on urban renewal programs in the South; however, studies of Richmond, Virginia, and Atlanta, Georgia, produced findings similar to those for Northern cities. These studies provided important information on Black input and participation in urban renewal programs. For example, in both cities unsuccessful protests were mounted in the 1950s against local proposals to raze parts of Black neighborhoods.[8] However, middle-class African Americans and representatives of Black institutions, such as the colleges and universities, the Urban League, and real estate companies, favored redevelopment and other proposals to improve living conditions in Black neighborhoods.[9] Urban renewal proposals put forward by White developers and public officials met the needs and desires of this segment of the Black population. In Richmond (though it was not the case in Atlanta), the city planning commission had one Black member who championed the cause of redevelopment.[10]

What is beginning to emerge from these more recent studies of urban renewal is a revisionist perspective on the impact of the government-sponsored redevelopment programs on urban Black populations. Whereas earlier researchers concluded that urban renewal should be considered a form of "Negro removal," studies conducted in the 1980s and 1990s suggest that the issue is more complex. While many lower-income Blacks were displaced, some middle-class Blacks welcomed the redevelopment programs and participated in the urban renewal decision-making process. As we shall see in Greensboro, North Carolina, the controversies generated by urban renewal proposals were even more complex because of the organized opposition to these plans coming from the Black business community.

THE CUMBERLAND PROJECT

By the early 1950s, some people considered Greensboro a model "New South" city because of the economic prosperity generated by a strong industrial base in textiles and the presence of several major insurance companies. It was considered cosmopolitan due to the presence of five colleges and liberal because of its support for racial tolerance—Greensboro was the first city in the South to announce its compliance with the 1954 Supreme Court's *Brown v. Board of Education* decision outlawing segregation in public schools.[11] To some extent, prosperity and sophistication were apparent in Black neighborhoods. Compared to other parts of the state, Blacks in Greensboro had higher median educational levels, and many were homeowners. In 1950, approximately 15 percent were employed in professional or related occupations.[12]

Of Greensboro's five colleges and universities, North Carolina Agricultural and Technical State University (A & T) and Bennett College were Black institutions. Serving as centers of social, cultural, and intellectual activity, A & T and Bennett helped to nurture the development of a sizable Black middle class. In Greensboro, Blacks had a long history of involvement with local politics and the struggle to overcome legalized segregation. Starting in 1933, Blacks ran for city office on a regular basis, and in 1951 the first Black was elected to the city council.[13]

This progressive tradition and Black political activism led to the appointment of Black educator and NAACP activist, Vance H. Chavis, to the Greensboro Redevelopment Commission (GRC) in 1953. This meant that from the beginning there would be some Black input in the urban redevelopment projects proposed and implemented in Greensboro. Vance Chavis was also one of the few Black members of the ADA (Americans for Democratic Action); it was after Mayor Robert Frazier had attended an ADA meeting and met Chavis that he asked him to become a member of the commission.[14]

The first program undertaken by the redevelopment commission was the Cumberland Project. Begun in 1959 in the East Market Street area, the Cumberland Project would profoundly change the context of Black life in Greensboro. "Market Street" had been the focal point of Black life from the 1930s. Blacks went there to have a tooth pulled by Dr. Barnes, to get a hot meal at Farley's Grill, to meet a sweetheart for a date at the all-Black Palace Theater, or to attend services at Trinity A.M.E. Zion Church, one of several prominent Black churches in the city. Unfortunately, the city's neglect of sanitation, building code enforcement, and infrastructural improvements during the postwar period led to the severe physical deterioration of Market Street, especially in the section referred to as the "Bull Pen," which was to become the site of the Cumberland Project.[15]

The GRC held its organizational meeting on February 26, 1952, electing David J. White chairman; Dr. Elizabeth D. Bridgers, vice chairman; and city planning head Ronald Scott as GRC secretary and acting executive director. Mayor Robert H. Frazier introduced Richard Bardolph, J. A. Dogget, and Vance H. Chavis as members of the GRC. The GRC designated East Market Street as Redevelopment Area no. 1; the city would condemn all buildings and transfer all inhabitants to more adequate housing facilities.[16]

Although it was announced that slum clearance would begin in February 1953, a protracted series of legal and legislative problems over the constitutionality of eminent domain delayed the project's implementation for almost seven years.[17] In September 1958, the GRC hired its first permanent executive director, Robert E. Barkley, who had been director of urban renewal for the Chattanooga (Tennessee) Housing Authority.[18] Barkley oversaw a survey of the Cumberland area's housing, which revealed the appalling extent of substandard conditions. For example, 15 percent of the dwellings lacked adequate toilet facilities, while 33 percent had toilets located only on back porches. Nevertheless, pockets of decent housing existed: 80 percent of the owner-occupied housing was up to standard in terms of facilities and maintenance.[19]

Residential displacement would become controversial in Greensboro, as it did in other cities. However, more significant was the struggle between public housing advocates and private real estate developers over the provision of relocation housing. Despite the opposition from the Greensboro Board of Realtors, Vance H. Chavis and other liberal GRC members wanted to ensure the construction of public housing for low-income Blacks facing displacement. Chavis and others faced major opposition because the antipublic housing forces had considerable resources and influence and had enjoyed many victories since the 1930s.[20]

From the enactment of the 1937 Housing Act, private builders and lenders nationwide had opposed public housing, alleging that government intervention had created undue competition in local housing markets. Opponents argued that public housing was expensive, its physical design fostered the creation of slums, and it was another example of government inefficiency. They asserted that the private sector could do a better job.[21] In contrast, the proponents of public housing distrusted the private developers' assurances that they could provide suitable, decent, and affordable housing for low-income individuals and families. The proponents argued that private realtors' primary concerns were profits and gaining urban renewal contracts and funding.[22]

As the GRC moved into the property acquisition stage of the Cumberland Project, it became apparent that replacement housing for 450 Black low-income families would be needed; unfortunately, Morningside and Ray Warren Homes, the city's two Black public housing projects, would not be able to fulfill this need, according to a survey presented at a GRC meeting held on January 13, 1960. This report was a response to the members of the Greensboro Board of Realtors who had argued that a significant number of families in the housing projects could afford private housing. Once these higher income families left, this would leave sufficient housing for those who were to be displaced. Using the survey results, Chavis estimated that even if the higher income families did move, this would create space for fewer than 100 dislocated families, leaving the remaining 250 families homeless. Commissioners Mark Arnold and Elizabeth Bridgers warned that public housing would have to be built or the GRC could not proceed with its plans.[23]

On April 19, 1961, GRC officials announced that the Federal Housing and Home Finance Agency (HHFA) would stop work on all renewal projects except for Cumberland, if assurances could not be given that the city would furnish 450 units of relocation housing for displaced families, for rents of $40 or less monthly. Such a decision would have halted construction on the Washington Project, the city's second redevelopment area, where the U.S. Post Office authorities had planned to build a $5,000,000 mail distribution facility.[24] The Washington Project represented the remainder of the East Market Street redevelopment program.

At the city council meeting held April 27, 1961, debate ensued over the adoption of the GRC resolution to provide public housing. R. E. Lowdermilk, president of the Greensboro Board of Realtors, said that his organization had adopted a resolution opposing the construction of an additional 450 public housing units; they still contended that the existing public housing projects would be sufficient if the higher-income families were forced out. Eight other realtors spoke out in opposition to building more public housing.[25]

Vance Chavis also testified before the city council regarding public housing. He stated that his organization, the Greensboro Men's Club, one of the oldest and most prestigious Black civic organizations in the city, had endorsed the building of public housing for poor Blacks likely to be displaced by redevelopment. Councilman E. R. Zane agreed with Chavis and went on to point out that public housing offered the only solution to the impasse. Despite the continued protest of about fifty realtors, on May 4, 1961, the city council unanimously adopted a resolution calling for 450 additional public housing units that would rent for no more than $40 a month. The resolution would help the city avoid a federally imposed order to halt plans for future renewal projects valued at more than $15,000,000.[26]

BLACK INPUT: PROPONENTS VERSUS PROTESTORS

The officials of various social institutions located in the blighted Cumberland community supported the redevelopment program from the beginning. Administrators and faculty of North Carolina A & T and Bennett favored redevelopment, as had middle-class Black residents in Atlanta and various Northern cities. The pastors of the most influential African American congregations in the city also embraced redevelopment as a cause. And it was at a public hearing on the Cumberland Project held by the GRC on August 18, 1959, that other members of the Black community made their support for redevelopment known. Approximately sixty Blacks attended the hearing, including representatives of Black colleges, the YMCA, professionals, as well as homeowners who were likely to benefit from urban renewal. Clyde Elrod, representing the Black Home Builders Association, praised slum clearance as the one of the finest programs ever developed, and he assured the GRC that he and his membership wholly supported the Cumberland plan. "The importance of the renewal plan in making the city more attractive and healthy," was explained by Dr. W. T. Gibbs, president of A & T. He pointed out that A & T's board of trustees had voted unanimously to approve the planned redevelopment and made clear its intention to buy land for the college in the Cumberland area after it was redeveloped.[27]

While many Blacks attending the hearing shared the opinion that the area was blighted, some were anxious and concerned about being uprooted and about the costs of moving and replacement housing. Nonetheless, they considered these problems secondary compared to the benefits of redevelopment. The Rev. H. W. Fields, pastor of Lindsay Street United Holiness Church, called the entire program "a godsend." To allay some fears over the situation for the displaced, Robert Barkley, GRC executive director, noted that studies had shown that the supply of homes and apartments would be adequate over the next two years. There would be almost 1,000 units available for 500 displaced families.[28]

As Cumberland redevelopment progressed, land became available for purchase. Those who expressed interest in buying in the area included a mix of Black institutions and middle-class professionals. For instance, at the March 13, 1962, meeting of the GRC, the Hayes-Taylor YMCA and A & T College offered to purchase the property located at the northeast corner of East Market and Dudley Streets. On June 12, 1962, GRC Director Barkley announced that bids were being accepted for single-family lots in Cumberland. Of those submitting bids, a majority came from Black politicians, educators, and other professionals.[29]

While a large portion of the Black middle class in Greensboro welcomed urban renewal, some professionals expressed misgivings about what redevelopment was doing to the Black community. On August 25, 1961, a brief article appeared in the *Greensboro Daily News* about a brochure that accused slum clearance of "killing Negro home ownership, business institutions and growth possibility." H. J. Kirksey, author of the twenty-four-page pamphlet, opposed locating the postal distribution center on the south side of East Market Street. Kirksey, a local Black publisher, argued that it was a traffic hazard to A & T and Bennett Colleges and that the cost of the proposed Cumberland Shopping Center had increased sharply from $380,000 to $520,000 over a nine-month period. The GRC was also criticized for its inability to provide employment opportunities for Blacks in the city. Kirksey stated that he was distributing the pamphlet primarily to members of the Greensboro Negro Merchants Association and the Warnersville Citizens Council.[30]

Black firms located in Cumberland were particularly imperiled by redevelopment activities. GRC Director Barkley said that only one third of the small businesses along East Market, which included those in Cumberland, could be accommodated in the proposed 3½ acre shopping center—the only property reserved for displaced firms.[31] Yet along the East Market Street corridor existed approximately seventy-nine businesses, including five professional offices. Of this total, ten were owned by Whites, the remainder by Blacks. Forty-two of these businesses were in the Cumberland area.[32] Surveys indicated considerable duplication among the firms with eleven cafes, six barber shops, and five beauty shops. Many were marginal operations, generating revenues between $3,000 and $6,000 annually.[33]

On June 6, 1961, a special GRC meeting was called to discuss the relocation of Black businesses out of the Cumberland area. At the outset Robert Barkley, using a map, explained why all the structures on both sides of East Market Street had to be demolished, which meant that all businesses along the street would be destroyed. The Black business owners contended that (1) Market Street did not require widening on both sides; (2) managerial and financial problems would keep almost 90 percent of Black business owners from relocating in the shopping center; (3) the consolidation of similar businesses that relocated into the shopping center was impractical; and (4) the south side of East Market between Bennett and A & T was an undesirable location for the future postal facility because it would create traffic congestion.[34]

Assistant City Planning Director Dennis Daye responded that widening of East Market Street was integral to the major thoroughfare construction planned for Greensboro. And as far as the location of the post office was concerned, M. A. Arnold, GRC vice chairman, added that this site had already received federal approval, thus it could not be changed unless the city wanted to forgo this opportunity to get the new postal facility and the employment it would create. The pastor of Grace United Holiness Temple, Elder Freeman, was anxious about relocation of his church, which was both a place of worship and a business. Barkley explained that the building was in the right-of-way on Pearson Street and could be taken by the city without any relocation assistance. He recommended the church split its functions and acquire both an industrial and commercial site for relocation.[35]

At this June 6 meeting, Vance Chavis offered a set of suggestions that he invited the GRC to explore:

- Widen only the north side of Market Street
- Retain standard business structures, not in the way of the postal facility and located on the south side of East Market Street
- Subdivide the Cumberland Shopping Center into smaller parcels of land

J. Kenneth Lee, a prominent Black attorney and businessman, made it clear that he represented a group of Black and White investors willing to develop the shopping center if the displaced business people were unable to do so. However, Mr. Lee also emphasized that his clients would not give priority to displaced business owners that wanted to relocate in the new shopping center.

According to the GRC's official minutes, the public hearing was orderly and smooth. However, a subsequent newspaper account indicated that heated outbursts punctuated the three-hour meeting. A. C. Bowling, president of Greensboro Negro Merchants Association, was quoted as threatening action: "If you don't do something to relieve this problem," he declared, "it is going to break out of hand. Redevelopment is pushing us too far . . . but we have ways of striking back . . . call it a threat if you like," said educator Ezell Blair Sr. At the hearing's conclusion, another meeting was set for July 11, when Black business leaders would present their relocation proposals.[36]

At the July 11 meeting, A. C. Bowling presented a resolution to save all the businesses on East Market Street. Bowling made it clear that his membership had little interest in the shopping center as proposed because of the requirement that each store had to offer a unique good or service. In response, lawyer J. Kenneth Lee presented his plan, which closely resembled the GRC's original proposal for the shopping center. Lee's plan called for a center composed of twenty-seven units. The rents would be comparable to those that business owners were paying currently for basically obsolete facilities. He estimated rents would range between $135 to $150 monthly. The center would offer air conditioning, off-street parking, and other modern amenities. It would contain a large supermarket, a variety store, and a movie theater. Lee added that it was not feasible to allow more than one of each type of business in the shopping center, if investors were expected to be interested in the project. Although he had not contacted any of the owners located in the Cumberland area, Lee declared that he would welcome any business that was financially sound into the center or as a stockholder in the development corporation.[37]

The Cumberland Project with its shopping center was completed in 1963. Gross expenditures for Cumberland were $4,633,000, and net income—the amount the city received after renewed property was sold—was $3,593,423. The city demolished a total of 501 structures of mixed or incompatible uses. A total of 242 families were relocated and 85 percent moved into adequate quality housing.[38] Many Black businesses were unable to locate in the new Cumberland Shopping Center, since the center could accommodate only twenty-six of the approximately seventy-nine Black firms along East Market Street. New commercial areas were built outside of Cumberland, but they provided space for no more than twenty-one firms.[39] According to the *Greensboro Record*, the government role in the project was officially ended on December 31, 1965, seven years after its beginning. In sum, no room was available for approximately one-third of all the displaced Black businesses.

It should also be noted that over the long term, the shopping center proved to be a major fiasco. It became a location for an assortment of small personal service and retail firms, including a beauty shop, laundromat, and a discount variety store, but most of the businesses were not economically viable and eventually closed down. The shopping center experienced an extremely high rate of business turnover and became desolate at night.

CONCLUSION

To determine the impact of urban renewal on the Black population in Greensboro, one must first recognize the different elements and interests within the Black community. The vast majority of those Black families that were displaced were able to relocate to more adequate and better quality housing. Some even relocated to the new public housing projects opened at the insistence of Vance Chavis and the other liberal members of the GRC. It is unclear whether the city council would have approved these measures if the federal government had not intervened and threatened to halt all urban redevelopment programs in the city. However, in addition to the 450 units of public housing, 95 units of low-income multifamily housing were built in the Cumberland area. Also, the city was able to locate ample land to accommodate 700 units of multifamily housing for Blacks displaced during urban renewal.[40]

The needs of the Black middle class who supported urban renewal were fulfilled, as they had been in Atlanta, Richmond, and other cities. This corresponds with the findings of other researchers who emphasized the diversity within the Black population and found that while poor and working-class Blacks were displaced, middle-class Blacks and their institutions supported urban redevelopment and received some benefits. In Greensboro, middle-class Blacks who favored the redevelopment of Cumberland got a renewed and improved community with a park and refurbished land on which to build new single-family housing. Administrators from A & T College were able to purchase land to build a new girls' dormitory. Planning officials responded favorably to the suggestions of the Black middle class because they corresponded with GRC overall plans.

In Greensboro, planning officials did not simply ignore Blacks who protested against renewal efforts and displacement. To be sure, Black business owners and Vance Chavis were unsuccessful in preventing the displacement of Black businesses or finding places for their relocation. However, they did offer suggestions to resolve these problems. Unfortunately, their recommendations were not accepted by the other members of the GRC.

No matter how you look at it, Greensboro's urban renewal program disrupted the Black business community. Redevelopment robbed East Market of its vitality and intensity. East Market went from an economically diverse center with almost 80 firms, to one with fewer than 30 viable businesses. It is quite reasonable to view this social and economic loss as part of the price the Black community paid for urban renewal.

This examination of the Cumberland Project in Greensboro provides additional support for revising traditional interpretations of the impact that redevelopment programs of the 1950s and 1960s had on urban Black communities. While many Black families were displaced, many benefited from improved housing in other sections of the city. Middle-class Blacks supported the renewal programs and many Black institutions purchased land in redeveloped areas. It was primarily the Black business community in the Cumberland area that was displaced and paid the highest price for the revitalization of the city. Future studies of the history of Blacks and urban renewal must take into account the diversity within Black communities in the 1950s and 1960s and the complexity of Black responses to the changing social and economic circumstances in twentieth-century urban America.

NOTES

1. Martin Anderson, *The Federal Bulldozer: A Critical Analysis of Urban Renewal, 1949-1962* (Cambridge, MA: MIT Press, 1964), 183-90.

2. Anderson, *Federal Bulldozer,* 65; Peter H. Rossi and Robert A. Dentler, *The Politics of Urban Renewal* (New York: Free Press, 1961), 224; Jon Teaford, *The Rough Road to Renaissance: Urban Revitalization in America, 1940-1985* (Baltimore: Johns Hopkins University Press, 1990), 158.

3. Teaford, *Rough Road;* Roy Lubove, *Twentieth-Century Pittsburgh: Government, Business, and Environment* (New York: John Wiley, 1969); Peter Medoff and Holly Sklar, *Streets of Hope: The Fall and Rise of an Urban Neighborhood* (Boston: South End Press, 1994).

4. Jane Jacobs, *The Death and Life of Great American Cities* (New York: Random House, 1961), 4-5; Chester Hartman, *The Transformation of San Francisco* (Totowa, NJ: Rowman and Allanheld, 1984), 323.

5. Teaford, *Rough Road,* 178-79.

6. Joel Schwartz, *The New York Approach: Robert Moses, Urban Liberals, and Redevelopment of the Inner City* (Columbus: Ohio State University Press, 1993), 188; Teaford, *Rough Road,* 178-79; Thomas O'Connor, *Building a New Boston: Politics and Urban Renewal, 1950-1970* (Boston: Northeastern University Press, 1993), 224.

7. Teaford, *Rough Road,* 178.

8. Christopher Silver, *Twentieth Century Richmond: Planning, Politics, and Race* (Knoxville: University of Tennessee Press, 1984), 219-22; Clarence Stone, *Economic Growth and Neighborhood Discontent: System Bias in the Urban Renewal Program in Atlanta* (Chapel Hill: University of North Carolina Press, 1976), 67-68.

9. Stone, *Economic Growth,* 85; Silver, *Twentieth Century Richmond,* 264-70.

10. Silver, *Twentieth Century Richmond,* 276, 280, 283. Silver does mention that the Black planning commissioner, Addison R. Cephaus, became a member of a progressive political slate called Richmond Forward in 1964.

11. William H. Chafe, *Civilities and Civil Rights: Greensboro, North Carolina, and the Black Struggle for Freedom* (Oxford: Oxford University Press, 1981), 4-6.

12. Ibid., 17.

13. Ibid., 24.

14. Vance H. Chavis, telephone conversation with author, September 11, 1995.

15. Danny H. Pogue, Alexander R. Stossen, and Ethel Taylor, "Market Street Forum Exhibit '81" (Greensboro, NC: Deal Publishing Company, 1981, 3-8. A catalog from the exhibit held from April 19 to May 3, 1981, at North Carolina A & T State University.

16. *Greensboro Record,* February 27, 1952, B1. Redevelopment Commission minutes, Redevelopment Commission Office, Greensboro, NC, March 3, 1953.

17. *Greensboro Daily News,* January 28, 1953, B1. Under the existing state policy of eminent domain, demolition could occur only if all the housing in the area was deteriorated. The presence of one good house could halt redevelopment activities. The Greensboro Redevelopment Commission wanted the 1951 state Urban Redevelopment Law amended to permit redevelopment if two-thirds of an area was substandard. *Greensboro Record,* April 27, 1953, B1.

18. Redevelopment Commission minutes, September 16, 1958.

19. *Greensboro Record,* December 2, 1958, B1; Redevelopment Commission minutes, December 12, 1958.

20. Catherine Bauer Wurster, "The Dreary Deadlock of Public Housing," in *Urban Housing*, ed. William L. C. Wheaton, Grace Milgram, and Margy Ellin Meyerson (New York: Free Press, 1966), 246-47; Martin Meyerson and Edward C. Banfield, *Politics, Planning, and the Public Interest: The Case of Public Housing in Chicago* (Glencoe, IL: Free Press, 1955), 24-25.

21. Wurster, "The Dreary Deadlock," 246-48; R. Allen Hays, *The Federal Government and Urban Housing: Ideology and Change in Public Policy* (Albany, NY: SUNY Press, 1985), 89-90.

22. Ibid.

23. *Greensboro Daily News*, January 13, 1960, B1.

24. *Greensboro Record*, April 19, 1961, B1. The postal facility, which would require the elimination of ORD Boulevard, was originally planned as a part of the Cumberland Project. However, because major opposition was encountered from public works, planning, and traffic department officials who objected to the closing of this major artery, the postal center was relocated to the Washington I Project. Redevelopment Commission meeting minutes, February 28, 1961, and March 6, 1966.

25. Ibid. The following also was done to comply with federal requirements to prevent work stoppage. City officials provided letters of agreement from the Greensboro Housing Authority and the commission to work together to ensure that displaced families be given first priority when public housing vacancies occurred. Also, Greensboro Mayor George H. Roach promised to create a Relocation Coordination Committee as soon as possible.

26. *Greensboro Record*, May 4, 1961, B1.

27. *Greensboro Daily News*, August 19, 1959, B1; Redevelopment Commission minutes, August 19, 1959.

28. Ibid.

29. Redevelopment Commission meeting minutes, March 13, 1962; June 12, 1962.

30. *Greensboro Daily News*, August 25, 1961, B1.

31. *Greensboro Record*, June 6, 1961, B1.

32. The remaining thirty-one black businesses along East Market Street were located in the Washington Redevelopment Area. There was a total of seventy-nine businesses along East Market: forty-two on the Cumberland side, thirty-one on the Washington side with about six other businesses scattered throughout the entire East Market Street area. *Greensboro Record*, June 6, 1961, B1.

33. Ibid.

34. Redevelopment Commission meeting minutes, June 6, 1961; *Greensboro Daily News*, June 7, 1961, B1.

35. Ibid.

36. Ibid.; *Greensboro Daily News*, June 6, 1961, B1.

37. Redevelopment Commission meeting minutes, July 11, 1961.

38. *Greensboro Record*, November 18, 1965, B1.

39. The shopping center was too small to accommodate all the businesses that wished to remain in Cumberland. Other sites became available, according to a September 2, 1961, article from the *Greensboro Record*. Black firms moved from the 700 and 900 blocks of East Market to stores newly built along the 1300 to 1500 blocks of East Market Street and the 900 block of Gorrell. Over time, twenty-one new stores locations were built.

40. *Greensboro Record*, June 6, 1961, B1.

CHAPTER 12

THE ROOTS AND ORIGINS OF
AFRICAN AMERICAN PLANNING IN
BIRMINGHAM, ALABAMA

CHARLES E. CONNERLY
BOBBY WILSON

In searching for an African American planning tradition in Birmingham, one must look to the neighborhoods where Blacks have lived.
As in other cities, what has been recognized as planning in the history
of Birmingham's development has been primarily the planning dominated by White, economic elite institutions, frequently with the intent to
limit opportunities for Blacks. Necessity forced Birmingham Blacks to
develop independent institutions that planned for improvements in
their neighborhoods. Operating either apart from each other or at odds
with each other, Birmingham's White and Black planning traditions
maintained a separate existence until 1974, when they became inter-
twined with the adoption of a citizen participation plan. That plan has
since become nationally recognized for its degree of citizen involvement
in the planning and community development process.[1] The adopted
citizen participation plan represented a triumph of African American

neighborhood-based planning over the top-down planning approach favored by the White planning regime because it placed significant responsibility for community development planning within the hands of Birmingham's neighborhood residents, both Black and White.

The purpose of this chapter is to trace the historic development of a Black planning tradition in Birmingham neighborhoods and to determine its influence on Birmingham's citizen participation plan. It will do so by sketching the historical role of civic leagues in Birmingham as neighborhood-based civic improvement organizations and by showing how the Black community planning tradition fostered by the civic leagues affected the development of the city's citizen participation plan.

HISTORICAL ROOTS OF AFRICAN AMERICAN PLANNING IN BIRMINGHAM

From the city's founding in 1871 to the 1970s, Birmingham's planning depended on economically privileged White institutions, both public and private. Ten investors, all White, incorporated Birmingham in 1871, after purchasing 4,457 acres of farmland where the South and North Alabama Railroad was expected to cross an existing railroad, the Alabama and Chattanooga. The area held rich deposits of coal and iron ore. By the end of the nineteenth century, the arrival of the railroad helped propel Birmingham to status as a major coal, iron, and steel center.[2]

Just as quickly, Birmingham's mining and industrial employment opportunities attracted large numbers of Blacks from the rural South. Birmingham industrialists felt that Black workers were superior to White workers for manual labor jobs, and industrial employment offered Blacks an alternative to sharecropping.[3] As a result, Birmingham quickly grew into a city with a large Black population. Forty-three percent of its population in 1890 was Black, and by 1930, its nearly 100,000 Black residents made Birmingham the eighth largest city in the nation in terms of Black population.[4]

By the turn of the twentieth century, the Tennessee Coal and Iron Company (TCI) had become the dominant firm in Birmingham. The U.S. Steel Corporation acquired TCI in 1907.[5] To expand its operations in the Birmingham district, in 1910 U.S. Steel contracted with a real estate development company headed by Birmingham developer Robert Jemison Jr. to construct the industrial new town of Corey (later Fairfield). Two years later, Graham Taylor wrote in *The Survey*, a Progressive Era social reform journal, "It is doubtful if in America there may be found a better planned industrial community."[6] But Jemison's plan did not include housing for low-income workers, many of whom were Black.[7]

In planning for the entire Birmingham area, one of the nation's early city plans, Warren H. Manning's 1919 *City Plan of Birmingham*, focused on the physical environment but made no mention of the housing and sanitary issues that affected the lives of both Black and White residents in Birmingham.[8] Only a few years earlier, *The Survey* had published a major study of Birmingham, with close attention paid to the housing and sanitation problems found in that city's poorer neighborhoods.[9]

In 1925, the firm of John Charles Olmsted and Frederick Law Olmsted Jr. prepared a study of parks and playgrounds in Birmingham, at a time when no publicly owned parks could be used by Blacks.[10] The following year, Birmingham instituted its first comprehensive zoning statute, one that separated White and Black residential areas.[11] In the 1930s, Birmingham was one of the first cities in the nation to obtain federal funds for subsidized housing, through the Public Works Administration housing program that preceded the public housing program. The first development in this program, Smithfield Court, was located in an area designated for Blacks, who were the only people who could live there.[12] The city's premier urban renewal project in the 1950s and 1960s, which resulted in the establishment of the University of Alabama at Birmingham, wiped out most of the city's largest Black neighborhood.[13]

Conditions in Birmingham's Black Neighborhoods

Within this context of planning neglect and malevolence toward Birmingham's Black residents, the city's Black neighborhoods frequently lacked basic public services and amenities. According to *The Survey*'s 1912 examination of health conditions in Birmingham, "While Birmingham has no tall tenements, she harbors an equally dangerous condition in the cottages of laboring people, particularly the Negroes."[14] According to Blaine Brownell's description of Birmingham in the 1920s,

It was in the city's "vacant spaces"—areas of undeveloped land bypassed for more pleasant sites by industry and White neighborhoods—that the majority of Birmingham's Negroes settled. These Black neighborhoods were generally situated along creek beds, railroad lines, or alleys, and they suffered from a lack of street lights, paved streets, sewers, and other city services.[15]

A 1937 report prepared by the Jefferson County Board of Health quantified the deficiencies in public services provided to Birmingham's Black neighborhoods. The report surveyed twenty-two blighted neighborhoods in Birmingham, of which thirteen were predominantly Black. At a time when Blacks constituted 38 percent of Birmingham's population, 70 percent of the dwelling units located in the twenty-two blighted neighborhoods were found in predominantly Black neighborhoods. Among the city's blighted neighborhoods, Black neighborhoods had two-thirds of the community toilets, 72 percent of the dwelling units lacking running water, and over 80 percent of the dwelling units using kerosene for lighting (see Table 12.1). Therefore the absence of basic public services, such as running water and electricity, was concentrated in the city's Black neighborhoods.[16]

Personal testimony also bears witness to the absence of basic public services in Birmingham's Black neighborhoods. Rosa Kent described her Collegeville neighborhood located in North Birmingham:

> Our community was a rejected community when it came down to the city. Every three or four houses didn't have any running water. The streets were so dirty, they weren't paved. Women had to hang out their clothes at night because if a car came through you couldn't see anything for the dust.[17]

Benjamin Greene moved to his Harriman Park neighborhood in 1948. On his street he found one water meter that served all the residents on that street, with the woman who owned the property served by the water meter charging all of the other residents for use of her water. Because the street was located on a hill, residents at the top of the hill had difficulty getting the water they needed. In addition, none of the roads in the neighborhood were surfaced and there were no sanitary sewers.[18] Similar circumstances existed in the Grasselli Heights neighborhood, located in southwest Birmingham. John Culpepper reports

Table 12.1 Distribution of Neighborhood Deficiencies, by Race Among Blighted Neighborhoods in Birmingham, 1933 (in percentages)

Predominant Race in Neighborhood	Black	White	Total
Number of dwelling units (DUs)	70.3	29.7	100
Number of DUs without running water	72.5	27.5	100
Number of DUs using kerosene for lighting	83.9	16.1	100
Number of community toilets	66.8	33.2	100

SOURCE: J. D. Dowling and F. A. Bean, *Final Statistical Report of Surveys of the Blighted Areas, Birmingham, Alabama, Showing Some Comparative Figures for the Years 1933 and 1935* (Jefferson County, Alabama Board of Health, 1937, 5-6).

that in the 1950s, when he moved into that neighborhood, the dust from the unpaved streets meant that a person could not sit in front of his house. Furthermore, the neighborhood had no fire hydrants.[19] Drainage could also be a problem in Black neighborhoods, with open ditches creating flooding problems after heavy rains. According to Simmie Lavender of the Jones Valley neighborhood, in his neighborhood, "they didn't have ditches, they had rivers."[20]

The Role of Civic Leagues in Addressing the Problems of Birmingham's Black Neighborhoods

It was problems such as these that provided a focus of activity for the civic leagues of Birmingham. Civic leagues were, and to a limited extent continue to be, organizations established in Black neighborhoods with the primary objective of improving the quality of life in the neighborhood.[21] Civic leagues existed in Black neighborhoods throughout Birmingham and Jefferson County; at least forty-one civic leagues existed in Birmingham between 1945 and 1975, and an additional eleven leagues were located in Jefferson County Black neighborhoods outside of Birmingham.

Unfortunately, little documentation exists on the origins of civic leagues, but the oldest known leagues originated in the 1930s.[22] In addition, the Alabama State Federation of Civic Leagues, an umbrella organization, began in 1933.[23]

Civic leagues were the primary organizations in the Black Birmingham community to define membership on the basis of location; unlike other organizations, they took in anyone who lived within a neighbor-hood.[24] Consequently, civic leagues operated within a set of specific neighborhood boundaries.[25] To obtain funds, the leagues assigned block or street captains to collect dues from people who lived in the neighbor-hood.[26]

The activities of the civic leagues primarily focused on neighborhood improvement, as the following examples illustrate:

• *Neighborhood beautification.* The Grasselli Heights Civic League worked to beautify the neighborhood by contacting owners who had not maintained vacant lots they owned. In cases where the lot owners did not cooperate, civic league members undertook the cleanup themselves, utilizing a heavy duty lawnmower purchased with civic league funds.[27]

• *Community centers.* The Harriman Park, Fountain Heights, and Brighton Civic Leagues each constructed community centers for their neighborhoods.[28] The Harriman Park Civic League raised $1,500 to purchase two lots in 1972 for the community center that was eventually built with Birmingham community development funds.[29]

• *Public safety.* The Grasselli Heights Civic League devoted its atten-tion to petitioning for the installation of fire plugs in that neighborhood.[30]

• *Recreation and parks.* To help address the lack of recreation opportu-nities, in 1973 the Woodland Park Civic League negotiated with the Birmingham School Board to use the playground facilities at the neigh-borhood school. This required the civic league to take out an insurance policy.[31] In the early 1940s, the founder and head of the Alabama Federa-tion of Civic Leagues, W. L. McAlpin, was instrumental in persuading the City of Birmingham to create the first city-owned park for Blacks.[32]

• *Water and sewers.* After the City of Birmingham annexed its neigh-borhood in 1949, the Harriman Park Civic League successfully peti-tioned city government to obtain suitable water mains, sanitary sewers, and a gas line.[33]

• *Street improvements and drainage.* In the late 1940s, the Riley Station Civic League initiated a fund-raising campaign for street and drainage improvements that raised nearly $4,200 through such means as direct solicitation of residents and a plate dinner sale.[34]

• *Zoning.* The Druid Hills-Norwood Civic League organized a suc-cessful petition campaign to get the neighborhood zoning changed from a multifamily to a single-family designation.[35]

- *Voter education.* In an effort to help Black voters obtain greater influence over local public decision making, civic leagues frequently trained residents in procedures of voter registration and voting. Under Alabama law, Black voter registration was particularly difficult prior to passage of the 1966 Voting Rights Act by the U.S. Congress.[36]

In general, therefore, the residents of Birmingham's Black neighborhoods responded to the lack of public services by creating their own neighborhood-based organizations to address these deficiencies. As a result, by the early 1970s, when Birmingham's traditional White and elite-dominated approach to planning came under question, the city's Black community was prepared to put forth an alternative, bottom-up approach that built on decades of experience at organizing Black neighborhoods for community improvement.

THE 1970s: THE ROLE OF THE BLACK COMMUNITY IN SHAPING BIRMINGHAM'S CITIZEN PARTICIPATION PLAN

The 1970s began with two non-Black organizations attempting to work with the civic leagues to improve the quality of life in Birmingham's Black neighborhoods. These two organizations had contrasting agendas, and the interaction between the two groups and Black neighborhoods would greatly influence the design of Birmingham's 1974 citizen participation plan.

The first of these groups was Operation New Birmingham (ONB) and its Community Affairs Committee (CAC). ONB's roots extend back to 1957 when the city's downtown and newspaper interests created a predecessor organization, the Birmingham Downtown Improvement Association. Renamed, ONB had become a public-private partnership by the late 1960s, representing significant private interests but located within city hall.[37] ONB formed its CAC in May 1969, in response to concern by Birmingham's Black leadership over issues of police brutality and employment discrimination and the failure of city government to respond to these conditions.[38] Its initial membership consisted of nine Black and nine White representatives, along with nine local public officials.[39] The CAC began its work in Birmingham's Black neighborhoods by responding to residents of Collegeville, a Black neighborhood located in North Birmingham, who had approached Birmingham's mayor and city council for help in addressing neighborhood crime. Working with Collegeville neighborhood leaders, CAC subcommittee members sought action from local government agencies to address a list of fourteen problems identified by the neighborhood.[40]

Building on this experience, in 1970 Operation New Birmingham, through the Birmingham Regional Planning Council, applied for and received U.S. Department of Housing and Urban Development (HUD) comprehensive planning assistance funds to hire staff and expand the Collegeville program to a broader Neighborhood Planning Program that would, by 1971, include sixteen other neighborhoods.[41] In each case, neighborhood representatives would meet with ONB staff members, prioritize problems, and then in conjunction with ONB staff members and CAC subcommittee members, meet with public officials to request actions to address these problems.[42] In many instances working with civic leagues, ONB's Neighborhood Planning Program essentially carried on the efforts of the civic leagues by soliciting improved public services from appropriate government agencies.[43] The major difference was that the civic leagues and other Black neighborhood groups now had the clout of ONB and the CAC behind them. According to an unnamed ONB official, public agency heads "realize CAC has as strong a representation as any group as far as influence is concerned."[44]

At about the same time that ONB was expanding its Neighborhood Planning Program, Greater Birmingham Ministries (GBM), a church-based, ecumenical organization, was beginning its efforts to achieve "systems change" in Birmingham's Black neighborhoods. In the early 1970s, GBM established the Block Partnership Program, which matched largely White, suburban churches with primarily Black neighborhoods in Birmingham. GBM often worked with neighborhood civic leagues, including groups in the same North Birmingham area where ONB had begun its Neighborhood Planning Program. In contrast to the downtown-oriented ONB, however, GBM's neighborhood initiatives were led by a United Methodist Church minister, George Quiggle, whose approach to community organization was influenced by the writings of Saul Alinsky as well as his own theological beliefs. The Block Partnership Program used resources from both the suburban church and the Black neighborhood to address needs identified by neighborhood residents.[45]

In the summer of 1973, HUD's Birmingham area director, Jon Will Pitts, wrote a memo to Birmingham Mayor George Seibels that would propel the city, ONB, GBM, and Birmingham's neighborhoods to redefine the role to be played by Black, as well as White, neighborhoods in shaping the city's planning and community development agenda. At this time, prior to passage of the 1974 Housing and Community Development Act, HUD was responsible for recertification of the Workable Programs, which were plans prepared by cities such as Birmingham to continue HUD assistance. Eight months earlier, in November 1972, responding to citizen protest over ONB's perceived role as an exclusive representative of community groups, Birmingham's city council had tried to clarify its contract with ONB so that it did not appear that ONB stood between citizens and city government.[46] Unsatisfied with this change, however, Pitts wrote Seibels in July 1973 that the city needed to revise its community development process so that citizens would have direct access to the Mayor's Office; so that they would be "integrally involved, on a continuous basis, in all phases of the community development process"; and so that citizen participation would be "under direct City Control."[47]

Since 1966, ONB's members, consisting of over 200 leaders from the private and public sectors, had served as the city's Citizens' Advisory Committee in compliance with HUD's Workable Program requirements.[48] HUD's requirement that citizen participation be under direct control of city government and that it provide direct access of citizens to city hall had the effect of placing ONB's role in citizen participation in doubt. HUD was also making clear that ONB's Neighborhood Planning Program was not a sufficient substitute for direct citizen involvement in community development. HUD Area Office Director Pitts wrote the Birmingham Regional Planning Commission, the agency through which ONB received HUD funds for the Neighborhood Planning Program, that the program was "more effective in dealing with complaints than in involving citizens in the community development process."[49]

As a result of Pitts's letter, the City of Birmingham was forced to revise its citizen participation program as well as to revise the role that ONB played in that program.[50] In response to HUD, the Birmingham's Community Development Department prepared a citizen participation plan under which citizens in each of Birmingham's ninety-one neighborhoods could participate in their area's Neighborhood Citizens Committee and elect officers for their committee. These officers would also serve as representatives to a Community Citizens Committee, which would represent each of the sixteen communities in Birmingham.[51]

Each of the sixteen communities would elect a chairman, who would sit on a citywide Citizens Advisory Board composed of twenty-five members. Birmingham's mayor would appoint the remaining nine members. Moreover, ONB would have "primary field responsibility for developing, organizing, and maintaining viable citizen participation groups in the neighborhoods and communities throughout Birmingham."[52]

Birmingham Mayor George Seibels publicly announced the citizen participation plan in January 1974 and instructed the city's Community Development Department to put it into effect immediately.[53] Although city council member David Vann proposed an alternative citizen participation plan, a council majority rejected his call for a public hearing on citizen participation in early February of 1974.[54] Soon thereafter, three Community Development Department staff members, along with three ONB staff members, began the process of taking the new citizen participation plan to the neighborhoods.

ONB staff ranked Birmingham communities by the order in which they should be contacted about the citizen participation plan and recommended that the citizen participation process begin in the predominantly White Woodlawn community, located on the city's east side. The Community Development Department's staff, however, selected North Birmingham as the inaugural community for the program, even though ONB had slated that community as the fourth one to be visited in the citizen participation process.[55] Chuck Lewis, a former Peace Corps volunteer who had written a research paper on citizen participation in the urban planning master's program at Virginia Polytechnic Institute, helped staff the Community Development Department's citizen participation effort. He was concerned that the mayor's citizen participation plan was too dependent on mayor-appointed nominees and that ONB participation would make it more difficult to sell citizen participation in Birmingham's Black neighborhoods. ONB had been linked with an effort to boost the White vote in the fall, 1973 elections and consequently Lewis believed that ONB's involvement in citizen participation would be suspect in the Black community.[56] As a result, he went to Rev. George Quiggle, whom he had first known when Lewis worked as a community organizer several years earlier in a Black public housing development. Lewis, knowing that Quiggle and GBM had many contacts in Birmingham's neighborhoods, told Quiggle of his concern about the mayor's citizen participation plan and asked for advice.[57] According to Quiggle, "Chuck didn't have to say much; we knew what to do."[58]

On February 28, 1974, Community Development Department and ONB staff held a meeting in the Black Collegeville neighborhood that was attended by twenty-one representatives of the neighborhood.[59] Among the neighborhood leaders present were individuals who had been local leaders in ONB's pilot Neighborhood Planning Program: Benjamin Greene, president of the Harriman Park Civic League, and Lula Menefee, president of the Collegeville-Harriman Park Coordinating Council.[60] At the meeting, the neighborhood representatives expressed concerns about a number of issues: (1) What would ONB's role be in the citizen participation plan? (2) Why did the mayor have the power to appoint one-third of the citywide Citizens Advisory Board? (3) Why not use existing groups to represent the neighborhoods?[61]

Clearly, the neighborhood leaders at the Collegeville meeting were not happy with the strong role to be played by ONB and the mayor in the citizen participation program and the relatively weak role to be played by existing neighborhood groups. These concerns were further expressed at a general meeting held in Collegeville on March 7, 1974, which was attended by about 125 residents of the Collegeville neighborhood.[62] A similar meeting was held that night in Fountain Heights, a Black neighborhood also located in North Birmingham. According to the *Birmingham News* account, a central theme expressed at both meetings was that the mayor's citizen participation plan assumed that the Black community was unorganized, thereby ignoring the community organizations already in place. According to this account, civic league presidents attended the meetings, one of them asking, "Why can't we use the organization we already have? Why do YOU have to organize us?" Harriman Park Civic League President Benjamin Greene said, "We don't want an extension of City Hall into our neighborhoods."[63]

The neighborhood residents at these March 7 meetings called for a public hearing on the mayor's citizen participation plan.[64] Five days later, Birmingham's city council, at the urging of Mayor Seibels, called for a public hearing on April 1, 1974.[65] Collegeville neighborhood leaders, including Benjamin Greene, Lula Menefee, and Rosa Kent (the latter held a number of neighborhood leadership positions, including presidency of the local Parents-Teachers Association), called on their neighbors to attend the April 1 public hearing.[66] In general, Black neighborhood leaders prepared for the public hearing by meeting at Evergreen Baptist Church. Assisting in the organization of these meetings were Rev. George Quiggle and Greater Birmingham Ministries.[67] GBM's organizing assistance had also been evident at the March 7 Collegeville and Fountain Heights meetings, through the presence of a spin-off organization, the Legal Evaluation and Action Project (LEAP).[68] At Evergreen Baptist Church, equipped with fifteen typewriters, Black neighborhood leaders planned their presentations for the April 1 public hearing.[69]

Because of the large turnout expected, the April 1 public hearing was moved from city council chambers to Birmingham's Boutwell Auditorium. Over 500 people attended the meeting, thirty of whom spoke. None of the thirty speakers favored the Mayor's citizen participation plan. Most of the speakers came from Birmingham's Black neighborhoods and at least eight were associated with various civic leagues.[70]

Generally, the speakers reiterated what had been said publicly at the meetings in the Collegeville and Fountain Heights neighborhoods. According to a Birmingham News article, "the most frequent complaint was that a City-sponsored network would displace already established and trusted organizations and might produce groups less responsive to needs of citizens now represented by organizations of their own choosing."[71] A tabulation of remarks made by Birmingham's Community Development Department showed that the second most frequently cited single concern was the impact that the mayor's citizen participation plan would have on existing organizations in the neighborhoods. The lack of citizen input was the most frequently cited concern. Other frequently cited concerns focused on the role of ONB, the impact that the three-level citizen participation plan would have on direct access to city government, concern with the mayor's authority to appoint nine members to the Citizens Advisory Board, and concern that traditional neighborhoods were not identified in the plan and were grouped with other neighborhoods on maps that Birmingham's planners prepared.[72]

Given the strong sentiments expressed at the April 1 public hearing, Birmingham's Community Development Department staff began to work on incorporating citizen concerns in a revised plan. Chuck Lewis and his staff organized a follow-up citizens meeting on April 25, 1974, to which they invited the thirty speakers at the April 1 public hearing and others. On the basis of comments made at the hearing, staff divided the approximately eighty individuals attending this follow-up meeting into five groups, each of which focused on one of the major issues identified at the public hearing. The Community Development Department staff then used the comments made at this meeting to revise the citizen participation plan.[73]

By June 27, 1974, the Community Development Department had revised the citizen participation plan, generally following the recommendations made by the participants in the April 1 and April 25 meetings. ONB's role in citizen participation had been jettisoned and, instead, Community Development Department staff would work with neighborhood groups. The new plan also called for community representatives to elect all of the citywide Citizens Advisory Board members; the mayor would no longer have the power to appoint any members. Existing neighborhood organizations, such as the civic leagues, were directly encouraged to participate in the citizen participation program. Finally, the Community Development Department staff pledged to work with local citizens to determine the names and boundaries of the city's neighborhoods, each of which, under the plan, would continue to have a Neighborhood Citizens Committee, with officers elected by neighborhood residents.[74]

With the citizen participation plan rewritten, the Birmingham Community Development Department staff completed the job they had previously begun of soliciting citizen opinion on the names and boundaries of Birmingham's neighborhoods. City staff used citizen interviews and met with neighborhood organizations to obtain information on residents' perceptions of neighborhood and community names and boundaries.[75] As a result of this effort, city staff developed a list of eighty-six neighborhoods, thirty-two of which were not on the list of ninety-one neighborhoods presented by the city in its initial citizen participation plan.[76]

The city staff's careful work in obtaining citizen input in delineating neighborhood boundaries helped to obtain support for the revised citizen participation plan. Tony Harrison, president of the Enon Ridge Civic League, said that with the revised neighborhood map, the city had, for the first time, placed on a map the oldest and most prestigious Black neighborhood in Birmingham—Enon Ridge.[77] Benjamin Greene, Harriman Park Civic League president, said that the revised citizen participation plan gave a neighborhood two things: (1) a voice in what would happen in the neighborhood and (2) an ability to identify its own borders.[78]

As a result, Black neighborhoods, as well as White neighborhoods, were to be recognized by the names and boundaries their residents ascribed to them. In addition to Enon Ridge, eleven other Black neighborhoods with Civic leagues that were not on the original citizen participation plan neighborhood map appeared on the revised neighborhood map.[79] Collegeville, which had appeared on the previous map as two neighborhoods, Douglasville and Hudson Park, appeared on the revised map with the boundaries and name known by the residents of that neighborhood.[80] In at least several instances, the neighborhood boundaries delineated under the revised citizen participation plan corresponded to the boundaries previously employed by the civic leagues in those neighborhoods.[81]

As a result of these changes, Birmingham's Black neighborhoods obtained the historical identities that their residents, often through the civic leagues, had given them. Of the forty-one civic leagues that have existed in Birmingham, twenty-nine have the same name as neighborhoods that are identified in the most recent Citizen Participation Plan for Birmingham neighborhoods.[82]

With these changes, Birmingham's Black community switched its position on the citizen participation plan from opposition to support. Birmingham's city council, however, had put the revised citizen participation plan on hold during the summer of 1974. Soon after the Housing Act of 1974 was signed into law in August 1974, three city council members attended a National League of Cities conference in Atlanta on the act and on the provisions of the new Community Development Block Grant (CDBG) program (see Part 5.G). Having been told that cities would need a citizen participation plan to participate in the CDBG program, the three council members recognized that the city had prepared a citizen participation plan, which simply needed to be adopted.

Within a week of their return to Birmingham, one of the council members attending the Atlanta meeting, David Vann, called for a public hearing to review the revised citizen participation plan.[83] At this public hearing, held October 1, 1974, the new plan was well received by the public; the city council ratified the plan one week later.[84] Several weeks after that, the eighty-six neighborhoods delineated in the Citizen Participation Plan each held elections for neighborhood association officers. In at least sixteen of these neighborhoods, elected officers included individuals who also participated in the neighborhood's civic leagues.[85]

CONCLUSION

The African American planning tradition in Birmingham was forced by the circumstances of discrimination experienced within the neighborhoods occupied by that city's Black residents. The roots of this tradition go back at least to the 1930s and consist of a rich network of neighborhood-based organizations that were formed to provide or petition for basic public services to the residents of their neighborhoods. By the 1970s, when the City of Birmingham was forced by HUD to reform its elite-dominated approach to planning, its attempts to retain a top-down, elite-controlled citizen participation process were thwarted by a Black community whose planning history was enriched by a tradition of neighborhood self-determination. While the actions of Birmingham's Black neighborhoods in shaping the 1974 citizen participation plan were aided by HUD's intervention, the city's Community Development Department's advocacy planning, and the community organizing efforts of Greater Birmingham Ministries, these forces from outside the Black community could not, by themselves, have created the firestorm of reaction in the spring of 1974 that resulted in the revision of the mayor's original plan. Instead, this energy and passion came from a community with a rich tradition of community planning that finally had the opportunity to impose its vision of planning on the larger community.

NOTES

1. Jeffrey M. Berry, Kent E. Portney, and Ken Thomson, *The Rebirth of Urban Democracy* (Washington, DC: The Brookings Institution, 1993), 301-3. Birmingham's Citizen Participation Plan was selected from over 900 nominations as one of five top citizen participation plans in the nation. Selection criteria included equality of opportunity for participation and ability of the citizen participation process to influence public policy.

2. H. M. Caldwell, *History of the Elyton Land Company and Birmingham, Alabama* (Birmingham: privately published, 1892), 3-4; Marjorie Longenecker White, *The Birmingham District: An Industrial History and Guide*, (Birmingham, AL: Birmingham Historical Society, 1981), 36-37; Martha Carolyn Mitchell, "Birmingham: Biography of a City of the New South" (Ph.D. diss., University of Chicago, 1946), 18.

3. Jonathan M. Wiener, *Social Origins of the New South* (Baton Rouge: Louisiana State University Press, 1978), 156-61; Carl V. Harris, *Political Power in Birmingham, 1871-1921* (Knoxville: University of Tennessee Press, 1977), 186-87.

4. Harris, *Political Power in Birmingham*, 34; Homer Hoyt, *The Structure and Growth of Residential Neighborhoods in American Cities* (Washington, DC: U.S. Federal Housing Administration, 1939), 64.

5. White, *The Birmingham District*, 92-94.

6. Graham Romeyn Taylor, "Birmingham's Civic Front," *The Survey*, January 6, 1912, 1467.

7. White, *The Birmingham District*, 122.

8. Warren H. Manning, *City Plan of Birmingham* (Birmingham, AL: City of Birmingham, 1919).

9. *The Survey*, January 6, 1912, 1449-1556.

10. Philip A. Morris and Marjorie Longenecker White, eds., *Designs on Birmingham: A Landscape History of a Southern City and Its Suburbs* (Birmingham, AL: Birmingham Historical Society, 1989), 51; Edward LaMonte, *Politics and Welfare in Birmingham: 1900-1975* (Tuscaloosa, AL: University of Alabama Press, 1995), 12.

11. *Birmingham Post*, July 13, 1926.

12. Federal Emergency Administration of Public Works, Housing Division, Statistical Section, "Status of PWA Housing Division Projects: December 15, 1936."

13. Charles E. Connerly, "Federal Urban Policy and the Birth of Democratic Planning in Birmingham, Alabama: 1949-1974," in *Planning the Twentieth Century American City*, ed. Mary Corbin Sies and Christopher Silver (Baltimore: Johns Hopkins University Press, 1996).

14. W. M. McGrath, "Conservation of Health," *The Survey*, January 6, 1912, 1508. McGrath, an engineer, was secretary of the Birmingham Associated Charities.

15. Blaine A. Brownell, "Birmingham, Alabama: New South City in the 1920s," *The Journal of Southern History* 38 (February 1972): 28-29.

16. J. D. Dowling and F. A. Bean, *Final Statistical Report of Surveys of the Blighted Areas, Birmingham, Alabama, Showing Some Comparative Figures for the Years, 1933 and 1935*, Jefferson County, Alabama Board of Health, 1937, 5-6; U.S. Bureau of the Census, *Fifteenth Census of the United States: 1930* (Washington, DC: U.S. Government Printing Office, 1932).

17. Interview with Rosa M. Kent, March 16, 1981, Birmingfind Project Interviews, file 809.2.4.2.5, Birmingham Public Library (BPL).

18. Interview with Benjamin Greene, former president, Harriman Park Civic League, June 23, 1995.

19. Interview with John Culpepper, former president, Grasselli Heights Civic League, June 27, 1995.

20. Interview with Simmie Lavender, Jones Valley Civic League, May 4, 1995.

21. Information on Birmingham civic leagues was obtained from interviews with twelve individuals directly involved in or knowledgeable of the civic leagues, a survey of the African American newspaper *Birmingham World* for the period 1945-1985, and examination of the uncataloged Emory Jackson files of the BPL. Emory Jackson was editor

of the *Birmingham World* from 1946 to 1975 and as such was a central repository for much information on the Black community in Birmingham.

22. The oldest civic leagues found in this study include: Ensley Civic League (1932), Brighton Colored Civic League (1934), South Elyton Civic League (1934).

23. "Thirty-Second Anniversary of Alabama State Federation of Civic Leagues, Inc.," December 5, 1965, Emory Jackson Papers, BPL.

24. Interview with Benjamin Greene, June 23, 1995.

25. Interviews with Benjamin Greene, June 23, 1995; Tommie Lee Houston, Grasselli Heights Civic League, June 23, 1995; Brazelia McCray, Hillman Civic League, June 24, 1995; and Calvin Haynes, president, Woodland Park Civic League, May 5, 1995; Constitution of the Woodland Park Civic League, n.d.

26. Interviews with Brazelia McCray, June 24, 1995, and John Culpepper, June 27, 1995; Constitution of the Woodland Park Civic League.

27. Interviews with Tommie Lee Houston, June 23, 1995, and John Culpepper, June 27, 1995.

28. Interview with Benjamin Greene, June 23, 1995; Program for Dedication of Fountain Heights Park Recreation Center, January 19, 1975, Emory Jackson Papers, BPL; "Brighton Colored Civic League," n.d., c. 1946, Emory Jackson Papers, BPL. Brighton is a suburb located just outside of Birmingham.

29. Resolution of the Harriman Park Civic League, November 15, 1978; Warranty Deed, November 8, 1972.

30. Interviews with Tommie Lee Houston, June 23, 1995, and John Culpepper, June 27, 1995.

31. Interview with Calvin Haynes, May 5, 1995.

32. Memorial Park was dedicated on May 31, 1942, *Birmingham World*, May 29, 1942. The second park for Blacks, dedicated several months later, was named for Mr. McAlpin.

33. Interviews with Benjamin Greene, February 24, 1995 and June 23, 1995.

34. *Birmingham World*, March 11, April 1, April 26, May 6, May 13, 1949.

35. Interview with Charlie Pierce, former president, Druid Hills-Norwood Civic League, June 24, 1995.

36. Interview with J. Mason Davis, former member, South Elyton Civic League, June 22, 1995.

37. Edward S. Lamonte, "Politics and Welfare in Birmingham, Alabama: 1900-1975" (Ph.D. diss. University of Chicago, 1976), 330-33.

38. Jim Murray, "Interracial Communication in Birmingham, Alabama and the Creation of Operation New Birmingham's Community Affairs Committee," 1990, typescript, 7-8.

39. Ibid., 8.

40. Birmingham Regional Planning Council, "An Application for Comprehensive Planning Assistance Funds from the Department of Housing and Urban Development," pp. V1-V4, May 29, 1970, Operation New Birmingham Papers, file 8.1, Department of Archives and Manuscripts, BPL.

41. Ibid., V4.

42. *Birmingham News*, June 3, 1971; Operation New Birmingham, "Neighborhood Planning Grant Project Completion Report, Phase I, August 1970—May 1971," July 9, 1971, 4-5, Operation New Birmingham Papers, file 8.3, BPL.

43. Operation New Birmingham, 1971, 6.

44. *Birmingham News*, June 3, 1971.

45. Interview with George Quiggle by C. Livermore, July 30, 1981; interview with George Quiggle by Connerly, June 7, 1993; Greater Birmingham Ministries, "Ministry Highlights in Greater Birmingham Ministries' History, 1968-1984," typescript, October 1984.

46. *Birmingham News*, November 8, 1972.

47. Jon Will Pitts to George Seibels, July 6, 1973, George Seibels Papers, file 49.34, BPL.

48. City of Birmingham, "Annual Review of Progress for Recertification of the City's Workable Program for Community Improvement," February 1, 1966, Albert Boutwell Papers, file 25.21, BPL.

49. Jon Will Pitts to William Bondarenko, August 3, 1973, Operation New Birmingham Papers, file 8.7, BPL.

50. James R. Land, Birmingham Community Development director, to William E. Ricker, executive director, Operation New Birmingham, August 10, 1973, George Seibels Papers, file 21.15, BPL.

51. Each community consisted of one or more neighborhoods.

52. City of Birmingham, "Proposed Citizen Participation Plan," November 1973, George Seibels Papers, file 18.53, BPL.

53. "Statement by Mayor George G. Seibels, Jr. Concerning New, Comprehensive Citizen Participation Plan," January 15, 1974, George Seibels Papers, file 18.53, BPL.

54. *Birmingham News*, February 5, 1974.

55. "Birmingham Citizen Participation Plan: Recommended Procedure for Executing the Field Responsibility of Operation New Birmingham," February 1974, Operation New Birmingham Papers, file 3.22, BPL; "Work Report, ONB Staff on Citizen Participation Plan, February 5-28, 1974," February 28, 1974, Operation New Birmingham Papers, file 3.22, BPL.

56. Interview with Chuck Lewis, February 23, 1995. ONB's connection to this effort was well publicized in the Black newspaper *Birmingham World*, November 3, December 8, 1973, as well as in the *Birmingham Post-Herald*, October 27, 29, November 27, 28, 29, 1973.

57. Interviews with Chuck Lewis, March 23, 1992, and February 23, 1995.

58. Interview with George Quiggle, June 7, 1993.

59. "Citizen Participation Plan Work Report: Collegeville Nucleus Meeting," February 28, 1974, Operation New Birmingham Papers, file 3.22, BPL.

60. Operation New Birmingham, "Neighborhood Planning Grant," 8.

61. "Citizen Participation Plan Work Report: Collegeville Nucleus Meeting."

62. "Citizen Participation Plan Work Report: Collegeville General Meeting," March 7, 1974, Operation New Birmingham Papers, file 3.22, BPL.

63. *Birmingham News*, March 10, 1974.

64. "Citizen Participation Plan Work Report: Collegeville General Meeting."

65. *Birmingham News*, March 12, 1974.

66. "Let's Put Action Where Our Mouth Is!," handbill, Operation New Birmingham Papers, file 3.22, BPL; "Citizen Participation Plan Work Report: Collegeville Nucleus Meeting."

67. Interview with George Quiggle by C. Livermore, July 30, 1981; interview with George Quiggle by Connerly, June 7, 1993.

68. *Birmingham News*, March 10, 1974; "Citizen Participation Plan Work Report: Collegeville General Meeting." Greater Birmington Ministries had organized and obtained funding for LEAP, which operated for four years in the mid-1970s. Greater Birmingham Ministries, "Ministry Highlights."

69. Interview with Benjamin Greene, March 24, 1992; interview with George Quiggle, June 7, 1993.

70. *Birmingham News,* April 2, 1974, April 24, 1974; *Birmingham World,* April 6, 1974; "People That Spoke at the Public Hearing, April 1, 1974," Operation New Birmingham Papers, file 3.22, BPL.

71. *Birmingham News,* April 2, 1974.

72. Ibid.; "Work Report, Citizen's Participation Plan, March 11-June 14, 1974," Operation New Birmingham Papers, file 3.22, BPL.

73. Interview with Chuck Lewis, February 23, 1995; *Birmingham News,* April 24, 1974, April 26, 1974; "Work Report, Citizen's Participation Plan, March 11-June 14, 1974," Operation New Birmingham Papers, file 3.22, BPL.

74. *Birmingham News,* June 27, 1974; City of Birmingham, "Citizen Participation Plan, October 1974," George Seibels Papers, file 18.53, BPL.

75. Interview with Chuck Lewis, February 23, 1995.

76. *Birmingham News,* November 20, 1974; City of Birmingham, "Proposed Citizen Participation Plan, November, 1973," George Seibels Papers, file 18.53, BPL.

77. Interview with Chuck Lewis, February 23, 1995.

78. Interview with Benjamin Greene, March 24, 1992.

79. *Birmingham News,* November 20, 1974. These neighborhoods are: Collegeville, East Thomas, Enon Ridge, Evergreen, Grasselli Heights, Harriman Park, Hillman, Jones Valley, Riley, Tarpley City, West Goldwire, Woodland Park.

80. "Work Report, ONB Staff on Citizen Participation Plan."

81. Interviews with Benjamin Greene, June 23, 1995; Brazelia McCray, June 24, 1995; Tommie Lee Houston, June 23, 1995; Calvin Haynes, May 5, 1995.

82. City of Birmingham, "Citizen Participation Plan: Neighborhood Associations: The Building Blocks of Birmingham," (Birmingham, AL: City of Birmingham, 1992).

83. Interview with Chuck Lewis, February 23, 1995; *Birmingham News,* September 10, 1974.

84. *Birmingham News,* October 8, 1974.

85. *Birmingham News,* November 20, 1974. The list of neighborhood association officers was surveyed for names of individuals who had been identified through interviews, examination of documents, and a survey of the *Birmingham World* for the period 1945-1985 as active members of civic leagues.

13

URBAN ENVIRONMENTALISM AND RACE

ROBERT W. COLLIN
ROBIN MORRIS COLLIN

The origins of White environmental consciousness are firmly embedded in a middle- and upper-class experience of nature, as reflected in the traditional conservationist movement in the United States. Staffing the movement are eager college students and college graduates who can afford to do nonpaid work. These students and graduates are predominantly White and middle or upper class. This important feeder mechanism for the U.S. environmental movement is lacking people of color.[1]

Separated both by class and race, and by the network of formal and informal segregation of housing, education, and employment documented in this book, African Americans and many other predominantly poor urban dwellers are not included in the predominantly White environmental movement. Yet most of the pollution is located in cities where the highest proportion of American poor and people of color reside.[2]

An impressive body of research indicates that environmentalists and minority urban dwellers are moving into an unavoidable relationship with each other. Environmental concerns are of great significance to urban and minority communities, and environmentalists are broadening their focus to include urban areas.[3] Regardless of the soil type, the hydrology of the bioregion, or the climatological conditions, it is the race of the humans who reside in the area that most determines the probability that a toxic or hazardous waste site is located there. For example, the greater the proportion of African American people in a community in the southeastern United States, the more likely that community is to be situated near a hazardous waste site.[4] In planning and public policy arenas, such findings demonstrate what is commonly known as environmental racism, or lack of environmental justice or environmental equity. Environmental equity is "premised on the notion of fairness in the distribution of environmental hazards, particularly those of a technological origin."[5]

As urban areas become more central to the goals of the environmental movement, the history of racial inequity and its inclusion as a factor in the discussion about environmental issues becomes critical. The provision of municipal environmental services such as recycling and basic sanitation is also tainted by the dynamics of racism, but we do not specifically address these issues in this chapter. Rather, we aim to give an overview of the development of "urban environmentalism" from before the 1970s until the present. As we will show, minority community activism has played an increasingly important role in environmentalism. Although this fact has caused some conflicts with traditional environmental activists, it has also redefined the environment agenda in creative ways. These include stronger attention to the immediate needs of oppressed populations and unique cooperative strategies between environmental justice advocates, industries, and government.

PRINCIPLES OF ENVIRONMENTAL JUSTICE UP TO 1980

A growing number of historians are widening their analysis of the early American environmental movement. Robert Gottlieb and others are challenging the view that early conservationists such as John Muir were the only contributors to that movement. Using a comprehensive interpretation of past environmental activities that includes early twentieth-century movements to clean up industrial pollution and improve sanitation and human health, these authors provide new insights into the roots of the more contemporary environmental justice movement in the last twenty-five years.[6]

Over the decades, many citizens whose voices either have not been heard or were silenced have sought to improve the quality of their everyday life. The particular way in which environmental movements view the differential impacts on people of color often depends on the context in which an issue develops. In the 1960s, for example, concerns about pesticides among mainstream environmentalists (as evidenced by the publication of Rachel Carson's *Silent Spring*) made it easier for the impact of such pesticides on Mexican farm labor to be "heard." To this day, the efforts of Cesar Chavez and the United Farm Workers to halt pesticide use has never successfully gained them access to pesticide policy debates.[7]

Both uranium mining and the evidence of lead poisoning in children are examples of environmental concerns shared by mainstream environmentalists and environmental justice advocates.

Since 1970, urban environmentalists have focused their attention on public health and working conditions. In the 1970s, events that had a race or class component may not have been identified as such, partly due to institutionalized racism in reporting practices of the mass media. However, urban environmentalists often sought to reform the environmental movement to include issues of race and class.

According to Michele Tingling-Clemmons, an African American activist,

> It was a revelation to the environmentalists that joining with workers could bring about these changes. I think this attitude plays a role in why environmentalism is perceived as something that is predominantly the concern of Caucasians who feel that they have a more global perspective than anybody else. This is a limited view on their part. It's limited based on their ignorance of history and on their sense of exaggerated self-importance.[8]

In 1976, a unique coalition of organizations formed to incorporate race and class issues into the environmental movement. This coalition sponsored important conferences in U.S. cities and formed a nonprofit organization called "Urban Environment Conference" (UEC). The UEC was an attempt by selected labor, environmental, and minority organizations to work together on urban environmental issues and to tap into the growing interest in the urban environment. The Sierra Club and some labor unions, in coalition, made the first organizational and financial contributions to UEC.[9]

A 1976 UEC conference in Michigan brought together representatives from labor unions, environmental groups, economic justice organizations, and churches to discuss issues of economic justice and the urban environment. The workshop topics indicated a full range of urban environmental concerns, from gardening to residential displacement and the impact of exclusionary zoning. As stated in the conference overview:

> City Care was built on the premise that it would be an open conference—structured to build mutual respect and confidence, and flexible enough to entertain ideas that might emerge. Plenary sessions heard major speeches on significant concerns, both national and grassroots. Workshops provided detailed information and discussion on topics that, while specific, enforced the concept of interrelationships integral to the City Care goal.[10]

A major result of this conference was a $66,000 federal grant to UEC, which the organization used to fund eleven similar "City Care" conferences throughout the United States.[11] Also helping to fund these conferences were several sponsoring organizations as well as federal agencies such as the Environmental Protection Agency (EPA) and the Departments of Housing and Urban Development, Interior, Energy, Agriculture, and Transportation. These conferences and other activities led to the adoption of the first "right-to-know" legislation in Delaware.

Community right-to-know laws state that communities should know whether hazardous materials exist in their midst. Originally passed in response to the concerns of fire officials and their unions, these laws spread rapidly across the United States during the 1970s at local, state, and federal levels of government, and they are central to locally based environmental activism. Whenever a waste site or industrial plant wants to open or expand in a given community, community right-to-know laws require notice of potential risks to the community. The knowledge of risk gained by affected communities empowered urban environmental activists and laid the groundwork for research and documentation of environmental injustices in the 1980s and 1990s.

One of the strongest federal community right-to-know laws is the Emergency Planning and Community Right to Know Act (EPCRA) of 1988,[12] which makes information available to community members about some of the toxic chemicals being handled that are specific to each community. Substantially supplemented in 1992 and in 1996, this law has its roots in the UEC conferences.

EPCRA has two basic aspects. First, EPCRA requires that business owners send Occupational Safety and Health data to the state emergency response commission, the local emergency planning committee, and the local fire department. Second, any business owner who is required to provide this information must also prepare an emergency and hazardous chemical inventory form and submit it to the appropriate state governmental body, local governmental body, and local fire department. The inventory form contains two types of information, Tier 1 and Tier 2. Tier 1 contains general information about the amount, location, and general characteristics of hazardous substances and is submitted yearly. It must be made available to any person requesting it. Tier 2 information contains more particularized information about the chemicals on the site of the business, including the name of the chemical and a description of the storage and location of the chemical. However, an owner of a business is *not* required to furnish this information unless the state, local government, or fire department requests it.

Once the state or local government requests this information and receives it, citizens have a right to request and receive that information as well. Thus, the community right to know mechanism in environmental law has become a fundamental part of urban environmentalism. Its success depends on the conscious involvement of state and local government officials, particularly environmental planners, who need to be activists in seeking local government cooperation.[13]

In 1979, in Detroit, Michigan, the National Urban League, the Sierra Club, and the UEC held a national conference of major significance. At this Conference on the Urban Environment, some 750 people, most of them community activists familiar with the grassroots impacts of national problems, convened for three and a half days. Addresses by leaders, information sharing, and the setting of new coalition agendas concerning the urban environment were the main focus of this conference. Issues addressed included public health, neighborhood empowerment, urban parks, suburban sprawl, and public transportation. Virtually all groups found common ground in these concerns. In the words of the conference organizers, "bolstered by the efforts in publicity and planning of some 75 national and community-based co-sponsoring organizations, the conference soon was recognized as part of an emerging process and not merely an important gathering."[14] Vernon E. Jordan Jr., then President of the National Urban League, stated,

> This is an important, perhaps historic conference. It marks the beginning of what we hope will be a new era of creative partnerships. People of all backgrounds and interests are gathered here out of a common concern for cities and for the people who live in them. By our presence we symbolically recognize the new realities of America's political and economic life, realities that demand cooperation in pursuit of change.[15]

In retrospect, these conferences were profoundly important first steps in building a coalition between environmental and civil rights movements. They symbolized the growing concern of African American groups for environmentally caused health hazards as well as the increasing focus of environmentalists on urban problems.

THE 1980s

By the early 1980s, African American environmental activists turned to direct protest to draw attention their issues. In 1982, African American people in Warren County, North Carolina, mounted one of the first national protests over the siting of a hazardous waste site. Their actions resulted in the arrests of 414 people. Governor Hunt had selected Afton, a small African American community in Warren County, as the disposal site for 32,000 cubic yards of dirt contaminated with toxic polychlorinated biphenyl (PCBs).

The hazardous waste was soil taken from 217 miles of North Carolina state highway, where Robert Burns and Ward Transfer Company had illegally dumped it. They were caught and criminally convicted for dumping PCB-contaminated oil along the state highway. The oil was transported from Buffalo, New York; rather than store it, Burns had taken the contaminated oil to the South and thrown it on the side of the road.

Afton was a stable community with a high home ownership rate and a high water table in porous, sandy soil. Warren County, which is rural, has the highest proportion of African American people and the lowest incomes in the state of North Carolina. Eighty-four percent of the residents in the area surrounding the site were African American, and 64 percent of the African Americans there owned their own homes. (The national home ownership rate for African Americans is about 42.5 percent; the overall national home ownership rate for all adults is about 64 percent.[16]) Most of the residents got their water from wells, and with the high water table and sandy, porous soil, many were concerned that PCBs leaking from the hazardous waste site would contaminate the drinking water. The residents signed petitions, held massive demonstrations, and picketed.

The protest received national publicity and became a turning point in the environmental justice movement. As Professor Robert Bullard noted, "Although the demonstrations in North Carolina were not successful in halting the landfill construction, the protests brought sharper focus to the convergence of civil rights and environmental rights and mobilized a nationally broad-based group to protest these inequities."[17]

Among those arrested in the Afton protests was District of Columbia Delegate Walter E. Fauntroy. Fauntroy persuaded the U.S. Government Accounting Office to study 1983 siting of hazardous waste landfills in EPA Region IV (Alabama, Florida, Georgia, Kentucky, Mississippi, North Carolina, South Carolina, and Tennessee). The resulting report, *Siting of Hazardous Waste Landfills and Their Correlation with Racial and Economic Status of Surrounding Communities*, concluded that African American people and ethnic minorities were more likely than the general population to live near a commercial waste treatment facility or uncontrolled waste site.[18]

This study spurred a national study by the United Church of Christ, led by Rev. Ben Chavis, who would later become the executive director of the NAACP. That study, *Toxic Wastes and Race in the United States: A National Report on the Racial and Socioeconomic Characteristics of Communities with Hazardous Waste Sites*, which was published in 1987, concluded that race was the best predictor of hazardous waste site location.[19] The Commission for Racial Justice employed Public Data Access, a New York-based research firm, to assist in the analysis of this information.

Statisticians poring over the results found only a 1 in 10,000 chance of such siting patterns occurring randomly.

At the same time, environmental groups who were studying the siting of incinerators reached the same conclusion. In 1990, Greenpeace published their report, *Playing with Fire*, in which they concluded that the percentage of minorities in communities with existing incinerators was 89 percent higher than the national median, and that communities with proposed incinerators had minority population levels 60 percent higher than the national median.[20] Together with these protests and reports, several judicial challenges to siting practices allegedly tainted by racism set the stage for the urban environmentalism of the 1990s.

THE COURTS AND ENVIRONMENTAL RACISM

Court cases focusing on environmental justice are rare because both poor communities and private corporations shy away from using legal means to solve problems in all but the most fierce disputes. Court cases are expensive, time-consuming, and sometimes disempowering. The following court cases are significant because they are representative of the deeper dynamics of urban environmentalism and environmental justice.

One of the first court cases was *Bean v. Southwestern Waste Management Corp.*, decided in 1979 in federal court.[21] Minority residents of Houston, Texas, opposed the siting of a hazardous waste site within 1,500 feet of a public school that had no air conditioning. They contended that they already had more than their fair share of hazardous waste sites. While the judge did acknowledge that the siting was "unfortunate and insensitive," he found that the neighborhood had not proven the necessary *racial animus* (racist intent) to prove their case. The available data were divided by census tracts; this diluted the neighborhood's case because the affected African American community and the school catchment area spanned several census tracts. This is a common problem in making a case.[22]

Another important case was *RISE v. Kay*, decided in federal court in 1991. RISE (Residents Involved in Saving the Environment) is a self-described biracial organization.[23] It brought suit because of the siting of a fourth regional landfill in King and Queen County, Virginia, which was

42 percent African American and 57 percent White. The judge considered the fact that three other local landfills were already located in the county, and all of them were sited in African American neighborhoods. The first (sited in 1969) was the Mascot landfill. All of the residents within a one-mile radius were African American. In addition, this landfill was located two miles from the Escobo Baptist Church, an important and historic African American church. The next sited landfill was the Dahlgren landfill (1971). Ninety-five percent of the people within a one-mile radius were African American. In 1977, the Owentown landfill was located where the population living in a one-half mile radius was 100 percent African American. The Owentown landfill was located one mile from another important and very historic African American church, the First Mount Olive Church.

Over the course of numerous public hearings, the residents voiced their concerns and offered alternative sites for the proposed new fourth landfill. They cited the traditional land use concerns of a decline in property values, the noise, and the risk to children that additional traffic would pose.

However, the judge held that the local government had adequately balanced its cultural, economic, and social needs in permitting a fourth landfill in an African American community. The judge did find a disproportionate impact by race but ruled that because no intent to discriminate was shown, it was therefore legal to develop the new landfill.

It is interesting to note that in 1986 a landfill was located in a White community in this same county. The community protested vehemently on the grounds of diminution of property vales, the noise, and the risk to children caused by the increased traffic. In contrast to *RISE v. Kay*, the county refused to permit this landfill to operate, the landfill owner lost every administrative and judicial appeal, and he had to close the landfill. The major difference between the fourth landfill challenged in RISE and all the others is that it is *regional*, not local. That means it brings in waste from other communities, waste not produced by the community being legally forced to take it. It also means that it is a much larger landfill, with much more traffic and much larger trucks. One cannot help wondering what sort of "balance" the judge struck in making this decision.[24]

Many communities of color individually struggled against such oppressive and racist land use practices through the 1970s and 1980s. When environmental justice became a national movement, all these struggling communities found new strength in a collective voice. A catalyst to the formation of this collective voice was the 1980 election of President Ronald Reagan, because Reagan did not cooperate with labor unions, and he did not make community concerns a priority in his administration.

Unfortunately, by the end of the 1980s, the fragile coalition between minority communities, labor, and environmental groups fell apart. Interest groups focused their efforts on their own constituencies and fought over what little federal funding and programmatic support remained during the administrations of Reagan (two terms) and Bush (one term beginning in 1988). The environmental problems of urban dwellers and African American communities escalated while the conduct and budgetary decisions of these two presidents subverted money and political will. As the pie got smaller, the divisions between groups increased, and the days of building coalition were, at least temporarily, over.

URBAN ENVIRONMENTALISM IN THE 1990s

The year 1991 brought the First National People of Color Environmental Leadership Summit, held in Washington, D.C. It was well attended by communities of color and environmental organizations and established a set of principles of environmental justice that eventually found their way into President Clinton's 1994 Executive Order on Environmental Equity. The Preamble developed at the summit states,

> WE, THE PEOPLE OF COLOR, gathered together at this multinational People of Color Environmental Leadership Summit, to begin to build a national and international movement of all peoples of color to fight the destruction and taking of our lands and communities, do hereby re-establish our spiritual interdependence to the sacredness of our Mother Earth; to respect and celebrate each of our cultures, languages and beliefs about the natural world and our roles in healing ourselves; to insure environmental justice; to promote economic alternatives that would contribute to the development of environmentally safe livelihoods; and, to secure our political, economic and cultural liberation that has been denied for over 500 years of colonization and oppression, resulting in the poisoning of our communities and land and the genocide of our peoples, do affirm and adopt these Principles of Environmental Justice.[25]

Three basic principles emerged from this summit. First, environmental justice is not separate from economic and political struggles in communities of color. Second, environmental justice is intrinsically related to community empowerment. And third, establishing a just environmental policy should be an inclusionary dialogue that recognizes that communities of color speak for themselves. The first principle was recognized in the 1980s but was not acted on by many government, environmental, or industry groups. While many gave community empowerment lip service in the 1970s and 1980s, not until this summit was this important principle driven home to powerful people in the environmental movement and in government. The summit also was the first gathering to recognize the necessity of a truly inclusionary dialogue:

> The event featured nearly 1,000 environmental justice activists from across the country, Africa, and South America, who met to develop a statement of principles and a call for national and worldwide action. Building on the momentum following the Summit, two groups formed: the Indigenous Environmental Network to promote the interests of Native American activists; and the Southern Organizing Committee, which organized the Southern Community Labor Conference for Environmental Justice in New Orleans in 1992, a gathering of nearly 2,000 activists from fourteen states.[26]

This summit alerted the federal government to the disproportionate impact of many environmental decisions and demanded some type of federal, especially EPA, response. In 1991 the EPA issued a groundbreaking report, *Environmental Equity: Reducing Risk for All Communities*, which underscored historical and present-day environmental inequities and focused on the need for governmental intervention where the pollution is the greatest—in communities of color.[27]

In February 1994, President Clinton signed Executive Order 12898 (see planning-related documents, Part 5.J, this volume).[28] This executive order on environmental equity shapes procedures in all federal regulatory agencies by ordering an analysis of disproportionate impact on communities of color for each suggested policy or action. Each federal agency is ordered to "make achieving environmental justice part of its mission by identifying and addressing, as appropriate, disproportionately high and adverse human health or environmental effects of its programs, policies, and activities on minority populations and low income populations."[29]

The executive order also created a federal task force on environmental justice. Members represent environmental justice advocates, other environmentalists, the EPA, labor, and industry. The task force made several concrete recommendations regarding procedures for evaluating environmental risk for projects and policies under review. These recommendations included evaluating current risk assessment methodologies, identifying communities particularly affected by environmental pollution in each EPA region, reviewing ways of improving environmental protection in Native American tribal lands, and supporting joint ventures for stakeholders, such as medical and public health initiatives. They also recommended implementing sustainable community strategies for affected communities and supporting students entering environmental justice work.[30]

One example of such an environmental evaluation is the measurement of risk in consumption of fish and wildlife specific to certain persons of color, such as Native Peoples in Michigan. A study by the University of Michigan revealed that some people of color rely on certain food sources much more than the majority of people in the United States and may consume fish in ways that could expose them to significantly higher risk.[31] Since one group therefore bears a higher level of risk, public health levels for "safe" consumption must reflect this reality rather than aggregate levels of consumption.

Moving away from setting standards reflective only of predominantly White and male norms is a critical and fundamental change in public policy, with major implications for urban environmentalism. Unfortunately, executive orders are not law per se but administrative guidelines. The next president could rescind or simply ignore them. However, Clinton's willingness to implement a more justice-oriented environmental policy may indicate an important step in the institutionalization of environmental justice.

Another significant effect of the executive order is the appointment of environmental justice representatives to the advisory bodies of federal agencies. The President's Council on Sustainability and the Common Sense Initiative are two examples of such advisory bodies. These advisory bodies are mandated to plan for our environmental future by making sustainability a priority in government planning and policy initiatives and to suggest innovative regulatory flexibility in order to decrease toxic emissions from industries.[32]

WORKING TOWARD A BETTER
URBAN ENVIRONMENTAL FUTURE

Research on cultural perceptions of the environment indicates that urban populations' perception of "environment" is more holistic, more inclusive, and more ecologically grounded than that of suburban populations.[33] In many communities of color, environmental problems manifest themselves as public health problems. Industry's pattern of externalizing the costs of wastes and environment toxins has impacted urban dwellers and African American communities more immediately than suburban dwellers. The health hazards resulting from such industrial pollution and toxins have contributed to the high death rate of African Americans at every age, with especially high death rates among African American infants and children. This has led urban environmentalists to organize around different rallying points and to seek fundamentally different types of solutions to environmental problems than do traditional environmentalists.[34]

To observe such environmental activism and problem solving at work (particularly in African American urban settings), one must examine public health initiatives launched by churches, mothers' groups, and other grassroots community organizations clustered around the visible health effects of toxicity in these communities. The organizing strategies these groups adopt resemble the strategies employed by the civil rights movement in the 1960s, strategies unlike the lobbying and litigation efforts of the mainstream conservation movement. The urban environmentalists also differ from traditional environmentalists in their approach to resolving environmental problems. Instead of seeking a perpetrator to punish through long-term policy change and/or litigation, urban environmentalists seek immediate reduction of risk in their communities.

These differences in approach are not necessarily incompatible; in fact, together they may lead to a more complete understanding of environmental problems and a more effective strategy to resolve these problems.

However, in the short term, they may lead to cultural miscommunication, differences in problem definition and perception, or raw political conflict. Industry may find a more ready partner in community groups seeking the immediate reduction of risk than in traditional environmental groups; traditional environmental groups may, on the other hand, suspect community groups of negotiating short-sighted solutions or forming ready coalitions with an "enemy." For example, the Common Sense Initiative discussed earlier has brought together environmental justice advocates, labor, state environmental regulators, the EPA, industry, and trade groups in six industrial sectors, in the first federally mandated consensus-building process in the United States. Other stakeholders must deal with environmental justice representatives to reach consensus.[35]

Coalition-building and dialogue among all the actors may bring into focus the single most important reality in the environmental debate: We share watersheds, airsheds, and, last, one world. While pollution and toxicity may be most easily forced on our weakest links—such as African American communities and urban places, because they are lacking in political power—the historical record shows that all of us ultimately bear the consequences.

CONCLUSION

The test of the maturity of the U.S. environmental movement as we approach the twenty-first century will be whether this movement can include communities of color in meaningful ways that involve sharing leadership and equality of project and policy control. If sustainability is the ecological goal of environmental public policy, then history shows us that dialogues including all stakeholders are necessary.[36]

The experience of the past twenty-five years provides a vast repertoire of strategies and modes of resolving complex, postindustrial environmental problems, including the challenge of bringing people widely separated by class, race, and world experience into a relationship of commonly held values. The methods of coalition-building from the conferences of the 1970s, the community organizing methods of the 1980s, and the sophisticated legislative and litigative initiatives of the contemporary era have the capacity to unite both urban and traditional environmentalists in a dialogue with industry and government. Escape from gridlocked, two-way dialogues between traditional environmental groups and industry is necessary in today's world. Both are now locked into a rhetoric of combat and fault that places communities and their immediate concerns at a distance.

The benefits that intense local involvement brings to the cause of environmentalism are clear; when local communities and industries buy into a solution to these problems, enforcement will be realistic, immediate, and inexpensive. Monitoring and enforcement mechanisms of a coercive, punitive government system are more expensive and less effective than the monitoring and enforcement mechanisms of an engaged citizenry. The combination of a comprehensive environmental movement with its eye on the global picture and community activism with its eye on local effects, especially on its most disenfranchised members, holds a promise for change in the destructive environmental practices of the industrial age.

NOTES

1. Dorceta Taylor, "Can the Environmental Movement Attract and Maintain the Support of Minorities?" in *Race and the Incidence of Environmental Hazards*, ed. Bunyan Bryant and Paul Mohai (Boulder, CO; Westview Press, 1992). In May 1991, The Wilderness Society had no people of color on its board, and minorities occupied only four of the eighty professional positions. The Sierra Club had one out of fifteen directors, the Audubon Society had two minorities out of thirty-three directors. Peter Steinhart, "What Can We Do about Environmental Racism?" *Audubon Magazine*, May 1991, 18-20.

2. Robert D. Bullard, *Dumping in Dixie: Race, Class and Environmental Quality* (Boulder, CO: Westview Press, 1990); Commission for Racial Justice, *Toxic Waste and Race in the United States: A National Report on the Racial and Socioeconomic Characteristics of Communities with Hazardous Waste Sites* (New York: Commission for Racial Justice, 1987).

3. Robert Collin, William Harris, and Timothy Beatley, "Environmental Racism: A Challenge to Community Development," *Journal of Black Studies* 25.3 (January 1995): 354-76; Robert Bullard, ed., *Unequal Protection: Environmental Justice and Communities of Color* (San Francisco: Sierra Club Books, 1994); Robert Bullard, ed., *Confronting Environmental Racism: Voices from the Grassroots* (Boston: South End Press, 1993); Bunyan Bryant and Paul Mohai, *Race and the Incidence of Environmental Hazards* (Boulder, CO: Westview Press, 1992); and U.S. Environmental Protection Agency, *Environmental Equity: Reducing Risk for All Communities* (Washington, DC: U.S. Government Printing Office, 1992).

4. See reports cited in notes 18, 19 and 20 below.

5. William M. Bowen, Mark J. Staling, Kingsley E. Haynes and Ellen J. Cyran, "Toward Environmental Justice: Spatial Equity in Ohio and Cleveland," *Annals of the Association of American Geographers* 85.4 (December 1995): 641.

6. Robert Gottlieb, *Forcing the Spring: The Transformation of the American Environmental Movement* (Washington, DC: Island Press, 1993), 29-30; Denis Binder, "Index of Environmental Justice Cases," *The Urban Lawyer* 27 (Winter 1995): 163-65.

7. Gottlieb, *Forcing the Spring*, 241-44.

8. Dana Alston, *Taking Back Our Lives: A Report to the Panos Institute on Environment, Community Development and Race in the United States* (Washington, DC: Panos Institute, 1990), 24.

9. Ibid.

10. Ibid., 5.

11. Over 200 people attended one of the last UEC events in New Orleans in 1983, "Taking Back Our Health—An Institute on Surviving the Toxic Threat to Minority Communities." The Reagan administration eliminated the federal agencies on which UEC depended (70-80 percent of its funding had come from federal sources).

12. 42 U.S.C. §§ 11021, 11044.

13. Collin, Harris, and Beatley, "Environmental Racism."

14. "City Care: Proceedings of a National Conference on the Urban Environment," held in Detroit, Michigan, April 8-11, 1979, 9.

15. Ibid.

16. *Statistical Abstract of the United States*, table 49, "Social and Economic Characteristics of White and Black Population, 1980-1994." This rate was down from 48.6 percent in 1980.

17. Bullard, *Dumping in Dixie*, 35.

18. United States General Accounting Office, *Siting of Hazardous Waste Landfills and Their Correlations with Racial and Economic Status of Surrounding Communities* (Washington, DC: U.S. Government Printing Office, 1983).

19. Commission for Racial Justice, United Churches of Christ, *Toxic Waste and Race in the United States: A National Report on the Racial and Socioeconomic Characteristics of Communities with Hazardous Waste Sites* (New York: United Churches of Christ, 1987), xiii-xv, 23-24.

20. Pat Costner and Joe Thornton, *Playing with Fire: Hazardous Waste Incineration: A Greenpeace Report* (Washington, DC: Greenpeace, 1990), 48-49.

21. 482 F. Supp. 673 (S.D. Tex. 1979), affirmed without opinion, 782 F. 2d 1038 (5th Cir. 1986).

22. Because census tract data often cross neighborhood lines, the impact of the facility can be diluted because of the number and distribution of people across multiple tracts. A planner or community activist or citizen interested in these issues will need to go to local definitions of neighborhoods and use block-level data to make a more solid argument. This is still no guarantee a given court will accept her reasoning, but it will provide more accurate evidence of the actual impact.

The most recent state study on environmental justice incorporated block data for the first time. See Washington State Department of Ecology, *A Study on Environmental Equity in Washington State* (Olympia, WA: Washington State Department of Ecology, 1995). For a discussion of the need to accurately measure environmental impact across geopolitical boundaries, see Robert Collin, "Environmental Equity and the Need for Government Intervention: Two Proposals," *Environment* 35.9 (November 1993): 41-43.

23. 768 F. Supp. 1141 (E.D. Va. 1991)(RISE I); 768 F. Supp. 1144 (E.D. Va. 1991) (RISE II).

24. For a more extensive discussion of the law, see Robert W. Collin, "Environmental Equity: A Law and Planning Approach to Environmental Racism," *Virginia Environmental Law Journal* 11.4 (Summer 1992): 495-546; Robert W. Collin, "A Review of the Legal Literature of Environmental Racism, Environmental Equity, and Environmental Justice," *The Journal of Environmental Law and Litigation* 9.1 (Summer 1994): 121-71.

25. Charles Lee, *The First National People of Color Environmental Leadership Summit* (New York: Commission for Racial Justice, United Church of Christ, 1992).

26. Deeohn Ferris and David Hahn-Baker, "Residential Segregation and Urban Quality of Life," in *Environmental Justice: Issues, Policies, and Solutions*, ed. Bunyan Bryant (Washington, DC: Island Press, 1995), 69.

27. U.S. Environmental Protection Agency report, *Environmental Equity: Reducing Risk for All Communities* (Washington, DC: U.S. Government Printing Office, 1993).

28. *Federal Register* 59 (1994): 7629, reprinted in 42 U.S.C. § 4321 (1994).

29. Ibid.

30. Ibid., s1-102.

31. Pat C. West, J. M. Fly, F. Larkin, and Robert Marans, "Minority Anglers and Toxic Fish Consumption: Evidence from a State-Wide Survey of Michigan," in *The Proceedings of the Michigan Conference on Race and Incidence of Environmental Hazards*, ed. Bunyan Bryant and Paul Mohai (Ann Arbor: University of Michigan Press, 1989).

32. The President's Council on Sustainable Development and the Common Sense Initiative are both federal advisory committees operating under the auspices of the Federal Advisory Committee Act. The Council on Sustainable Development has just completed its work on phase one by issuing a report to President Clinton titled *Sustainable America*, available from the Government Printing Office. The Common Sense Initiative has been functioning through the Environmental Protection Agency (EPA), reporting through EPA administrator Carol Browner, and formulating projects and policy recommendations for more effective enforcement of the environmental protection laws and regulations.

33. Rachel Kaplan and Janet Frey Talbot, "Ethnicity and Preference for Natural Settings: A Review and Recent Findings," *Journal of Landscape Architecture and Urban Planning* 15 (1988): 107-14, 112.

34. Carl Anthony, "Why African Americans Should Be Environmentalists," *Race, Poverty and Environment* (April 1990): 5-6.

35. Both authors are Environmental Justice representatives in the printing industrial sector and active in the Common Sense Initiative processes.

36. Robin Morris Collin and Robert W. Collin, "Where Did All the Blue Skies Go? Sustainability and Equity: The New Paradigm," *The Journal of Environmental Law and Litigation* 9.2 (1994): 399-460.

PART FOUR

PLANNING EDUCATION

See Part 5.I, for the
American Institute of
Certified Planners'
"Code of Ethics."

14

THE STATUS OF PLANNING
EDUCATION AT HISTORICALLY
BLACK COLLEGES AND UNIVERSITIES

The Case of Morgan State University

SIDDHARTHA SEN

Much has been written about the history of planning education at majority institutions.[1] Such an effort has not generally been undertaken for historically Black colleges and universities (HBCUs).[2] The lack of literature on the history of planning and related programs and planning education at HBCUs is perhaps attributable to two factors. First, accredited graduate programs in planning are rare at HBCUs. Second, heavy teaching loads and little support for research at such institutions have left little time for scholarly work.[3] This chapter explains why and how the first accredited planning program developed at Morgan State University, one of the nation's oldest HBCU institutions, despite the general absence of such programs at HBCUs. It explores the unique role of an HBCU planning program in increasing the representation of African Americans in the profession and the unique challenges facing planning programs at HBCUs.

EVOLUTION AND GOALS OF HBCUs: AN OVERVIEW

HBCUs are generally defined as Black institutions of higher learning established prior to 1964 with the primary goal of educating African Americans. HBCUs must be accredited by a nationally recognized accrediting agency or must be making an effort to get accreditation.[4] However, certain institutions that were established after 1964 are still designated as HBCUs by the National Association for Equal Opportunity in Higher Education (NAFEO). This section briefly describes the salient features of HBCUs from 1865 to the present.[5]

Both private and public HBCUs arose during the post-Civil War period (1865-1895). HBCUs were primarily a way for African Americans to gain higher education in a society that restricted access to college education either by law, in the South, or by social norms, elsewhere in the United States. Although many of them had such titles as "colleges" and "universities," in reality they were mainly elementary and secondary schools. Many of them became colleges for Black clergy and then teachers as the demand for African American teachers increased. Grounding students with "Christian education" was the primary goal of the private HBCUs. The second Morrill Act of August 1890 paved the way for public HBCUs, created by Southern and border states mainly to prevent African Americans from attending White land grant colleges and to limit their education to vocational training. The legacy of industrial, agricultural, mechanical, and vocational education for African Americans began during this period.[6] Liberal arts education was available at only a handful of private HBCUs.

In the period of legal segregation of White and Black colleges from 1886 to 1953, HBCUs were the prime suppliers of African American teachers for Black schools.[7] By the 1910s, secular philanthropic organizations became more prominent than Northern missionaries in nurturing African American education. However, most of these organizations had to assure the Whites in the South that the thrust of educating Blacks would be industrial and vocational training, since other professional positions were not open to African Americans. By the 1930s, most HBCUs had developed into full-fledged colleges, increasing graduate work and requiring a high school diploma for student entrance. In the 1930s, the National Association for the Advancement of Colored People (NAACP) and the NAACP Legal Defense and Educational Fund (NAACPLDEF) carried out a desegregation campaign that almost ended the "separate but equal" doctrine. By 1931, the Southern Association of Collegiate Schools (SACS) agreed to establish procedures to accredit HBCUs, and the American Medical Association approved thirty-one Black institutions for premedical courses. Another milestone of this period was the establishment of the United Negro College Fund (UNCF) in the 1940s to pool the resources of private HBCUs, which were going through a financial crisis.

On May 17, 1954, in *Brown v. Board of Education,* the U.S. Supreme Court ruled that "separate but equal" education was unconstitutional. Title VI of the Civil Rights Act of 1964 prohibited the spending of federal funds in segregated colleges. The Higher Education Act of 1965 made a wide variety of financial aid programs available to Black students; this increased minority enrollment in U.S. colleges and universities. Title III of the act signaled a federal commitment to survival and enhancement of HBCUs. In 1970, the NAACPLDEF brought a lawsuit (*Adams v. Richardson*) against the Department of Health, Education, and Welfare for not enforcing Title VI of the Civil Rights Act against states that operated dual segregated systems of public higher education.[8] In its ruling in the *Adams* case, the Court mandated the enforcement of deseg-regation laws and stipulated that the states must achieve a better racial mix of students, faculty, and staff. At the same time, the Court acknowl-edged the role of HBCUs in providing greater access and retention of African Americans and stated that White institutions should not recruit African Americans at the expense of HBCUs.

Several events of the 1970s and 1980s shifted U.S. educational policy from integration to preservation of HBCUs. After the Adams case, many HBCU officials were concerned that desegregation would be achieved by closing HBCUs or merging them with majority institutions, but this did not happen. First, HBCUs were graduating African Americans in underrepresented professions in significant numbers. Second, racism faced by African Americans enrolled in White colleges in the 1970s and

1980s established the viability of the HBCUs' role in nurturing Black students. Third, civil rights groups brought lawsuits to enforce rulings in the *Adams* case, which prevented HBCU mergers with majority institutions, and federal policies became supportive of HBCUs.

Despite these improvements and increased federal support, several setbacks hindered HBCUs as viable educational entities, including the Court dismissal of the *Adams* case in 1987 and requirements that the Department of Education monitor desegregation plans for states that operated racially segregated universities.[9]

PLANNING AND
RELATED PROGRAMS AT HBCUs

The Southern location and agricultural roots of HBCUs may have hindered the development of interest in urban problems.[10] Other reasons include the low popularity of planning as a field of education among African Americans, the absence of role models, and the lack of job opportunities in planning and allied professions.

An African American architectural educator, Professor Anthony N. Johns Jr., recalls that he grew up in a town where no one had ever heard of a Black architect or planner. In the 1950s, virtually no HBCUs offered such disciplines. African Americans were not employed in public or private development projects, he notes, even in the early 1960s. It was the War on Poverty programs and the Great Society years that opened doors for African Americans in such professions. Elva E. Tillman, an African American planner and educator, recalls that when she started working for the planning department in the City of Baltimore in the early 1970s, only a few Blacks worked in the department. It was such a novelty to hire African Americans that the planning department would make an effort to feature them in newspaper photographs.

Because education was traditionally the primary graduate degree attraction for African Americans and was one of the few "open" professions, it provided many role models.[11] In contrast, only since 1981 has the attractiveness of other majors such as business and management, engineering, and health-related professions increased among African

Americans.[12] However, urban planning has never been a popular field of study among African Americans.

Black architectural educators started a movement in the late 1960s and early 1970s to establish planning education at selected urban HBCUs.[13] They realized that the problem of too few African American planners was in part due to the absence of planning programs at HBCUs. Another rationale for such schools was for African Americans to themselves develop an urban theory of empowerment. It was further argued that the planning profession, never attuned to the reality of Blacks, could never alleviate the problems of African Americans in the inner cities.

Given the absence of planning schools, prominent HBCU architecture departments took up leadership for establishing planning programs at HBCUs. An Ad Hoc Council of Black Architectural Schools, set up in 1969, broadened their concerns to include skills and disciplines such as urban design and city and regional planning. The council included representatives from architecture departments of Hampton Institute (now Hampton University), Howard University, Tuskegee Institute (now Tuskegee University), Southern University at Baton Rouge, Tennessee Agricultural and State University at Nashville (which merged with the University of Tennessee and is now Tennessee State University), Prairie View Agricultural and Mechanical University, and North Carolina Agricultural and Technical State University. The council wanted to expand to include HBCUs without architecture programs, such as Shaw University, the six Atlanta University Center institutions, Fisk University, Virginia Union University, and other urban HBCUs with similar reputations and potential. But the American Institute of Architects, which had considerable control on the council, was opposed to including nonarchitectural departments.[14]

The initiative never materialized in Black architecture schools at the graduate level.[15] Among the architecture schools just listed, only Howard University's planning program retained a good reputation and a viable existence for a significant period of time. This program too was eliminated because of internal conflicts, lack of funds, and mismanagement.

In the early 1970s, faculty who had obtained architectural and urban planning degrees from institutions such as Harvard and MIT led a movement to create a School of Architecture and Planning at Howard.

According to Johns, this was an indication of how HBCUs emulated "models" from majority schools. This was a different view from that of Melvin L. Mitchell and other Black coworkers, who argued for setting up alternative planning programs at HBCUs.[16]

The School of Architecture and Planning at Howard began in 1973-74 with separate departments of architecture and planning. The faculty obtained a grant from the Ford Foundation to start the planning program. The School of Architecture and Planning evolved from a small department in the School of Engineering to a full-fledged school. Johns strongly believed that such an immense transformation could not have taken place at other HBCUs, because many of these schools did not have Howard's resources, particularly its large pool of architecture faculty with training in planning.

Johns hypothesized that "urban studies" programs with fewer resources arose at HBCUs primarily in response to the need for "quasi-planners" for antipoverty programs under the Kennedy and Johnson administrations.[17] This led to the growth of urban studies/urban affairs programs in the 1960s and 1970s, including the ones at HBCUs, and to programs and proposals designed to train quasi-planners at junior colleges to meet the shortage of professional planners.[18] Even now, a significant number of HBCU degrees awarded at the graduate level fall under the broad rubric of "public affairs and services."[19]

By the late 1960s, it became obvious that Black studies programs—forerunners of the urban studies programs—had failed in their intended goal to equip African American youth with the skills needed to gain metropolitan economic parity.[20] Urban studies programs replaced such programs and aimed in part to educate African Americans as lower-level social control technicians able to maintain social services in the "ghettos." Clearly, the demand for African American quasi-planners had an effect on the growth of urban studies programs at HBCUs.

The number of urban programs at HBCUs rose from twelve to twenty between 1973 and 1983.[21] However, most of these programs were in public affairs, urban studies, urban politics, and public administration. Only Alabama Agricultural and Mechanical University, Federal City College (now the University of the District of Columbia), Howard University, and Morgan State College (now Morgan State University) had professional planning programs.

MORGAN STATE UNIVERSITY AND
ITS COMMITMENT TO URBAN PROGRAMS

Morgan State University was founded in 1867 as the Centenary Biblical Institute, set up by missionaries for the sole mission of training African American men for the Methodist ministry.[22] The Centenary Biblical Institute was private, governed by a five-member board of trustees composed primarily of ministers.

The change to Morgan College occurred in 1890 and brought a broadening of the mission to educate men and women for careers other than ministry. The primary mission was to prepare African Americans of good moral standing for careers in public school teaching. It was still a private and church-controlled institution, but its governing body was expanded to include prominent citizens of the community.

In 1939, Morgan College was purchased from the Methodist Episcopal Church by the State of Maryland, which set up an independent board of trustees that governed Morgan College from 1939 to 1967. From 1967 to 1975, the State Board of Trustees of State Colleges governed Morgan. As Morgan College began to mature, its mission expanded from teacher training to liberal arts education. In 1975, the Maryland General Assembly granted university status to Morgan. This university status came with a new mandate to address and resolve urban problems and to become a multicultural, comprehensive urban institution. The Higher Education Reorganization Act of 1988 reaffirmed Morgan's status as an independently governed public urban university. In fact, it is the only state-supported institution in Maryland designated as an urban university .

In its transition from an HBCU to a multiracial university, Morgan retained its historical commitment of training African Americans and admitting students who were considered underprepared to do collegiate work but had the potential to succeed with proper guidance. Morgan's contemporary period brought new professional schools, emergence of the College of Arts and Sciences as the center of Morgan's academic community, and the development and reorganization of programs to meet the needs of today's urban populations. Currently, Morgan State University is a comprehensive urban-oriented institution with programs leading to undergraduate liberal arts, preprofessional, professional, masters, and doctoral degrees. A Board of Regents nominated by the

Governor of Maryland and approved by the State Senate governs the university.

HISTORY OF PLANNING-RELATED
EDUCATION AT MORGAN

Planning and related programs at Morgan[23] date back to the early 1960s.[24] The seeds of the planning programs developed in 1956 with the arrival of a young African American faculty member, Homer E. Favor. His dissertation, on property value and race, aroused his interest in poor urban communities. From 1956 onward, he became actively involved in community planning activities in Baltimore neighborhoods.

In 1963, the president of Morgan State College, Martin D. Jenkins, received a grant to carry out extension work in Baltimore's neighborhoods. Jenkins wanted to bring in a prominent Black sociologist from Bryn Mawr to carry out Morgan's community extension efforts, but this eminent sociologist was unable to accept the offer. Jenkins asked Favor to manage the program for a year.

The socioeconomic and political changes in the United States at that juncture of history influenced Jenkins' effort to carry out community development. Jenkins was sensitive to the "urban crisis," broadly defined as racial polarization, class alienation, physical decay, and deterioration of race relations in inner cities.[25] He was also aware of the War on Poverty programs and concerned about the university's role.[26] He felt strongly that, as a Black institution of higher learning, Morgan should be a pioneer in alleviating the urban crisis.

The grant led to the formation of the Urban Studies Institute, which was established in 1963 and focused on research and extension in the community. The institute's urban extension efforts included organizing problem-oriented conferences and educational programs for the communities. The conferences were considered to be a phenomenal success for two reasons. First, this was the first instance of a large number of Whites coming to Morgan's campus, a significant achievement in itself since Morgan had a "reputation" as a campus to which Whites would not come. Second, nationally known people came to these conferences. Much of this was attributable to Favor's reputation in the Baltimore area and the nation and to his personal connections. Impressed with the Institute's first year, Jenkins asked Favor to stay as director.

In 1970, Jenkins' personal connections at the Ford Foundation, and his commitment to planning-related programs, were important factors in the development of "the Center for Urban Affairs." Jenkins gave Favor the responsibility for writing the proposal, which encompassed six extension and instructional programs. The State of Maryland and the Ford Foundation funded the proposal, and the Center for Urban Affairs came into existence in 1970. The state eventually supported all programs after the Ford Foundation grant expired.

Institutional changes at Morgan influenced the administration to undertake this endeavor. When the proposal for the center was finalized in 1970, the state was considering a recommendation to develop Morgan State College as a racially integrated, urban-oriented university. This expanded urban mission prompted the administration to nurture the idea of a Center for Urban Affairs. In his preface to the grant application, Jenkins notes, "The Center for Urban Affairs is the first phase of the envisioned expanded urban thrust of the College."[27] The emerging role of the university in solving the urban crisis also made programs such as urban studies and urban affairs very popular in the 1960s and 1970s. Morgan's administration, faculty, student body, alumni, and supporters shared consensus that Morgan should play a greater role in alleviating the urban problems of the Baltimore metropolitan area and the state of Maryland.

The institutional culture at Morgan further facilitated the development of the Center and its programs. As an HBCU, Morgan was devoted to educating disadvantaged African Americans. It had an institutional culture of reaching out to underprepared students from inner cities, it was located in an urban area, and it already had a number of urban-related programs, courses, and research activities.

The undergraduate program in urban studies was specifically designed to meet the institutional changes facing HBCUs, and Morgan in particular, by preparing students for service in the African American community, participation in the larger society, and a role in alleviating urban-oriented problems. The urban studies program was to provide a basic understanding of the political, social, spatial, and economic components of contemporary urban society.

External support facilitated the creation of the instructional programs. The Baltimore area chapter of the American Institute of Planners (AIP) endorsed the idea of undergraduate urban studies and graduate study leading to a master's in Urban Planning in a resolution passed in October 1969. The resolution also called on the national chapter of the AIP to foster the development of the graduate program. No graduate-level program in planning existed in the Baltimore region, or in the state, at that time. Undergraduate quasi-planners were needed to work at various public and quasi-public agencies.

Morgan initiated the M.A. in Urban Planning and Policy Analysis in 1970 and the undergraduate program in Urban Studies and Community Service in 1971. During its initial years, the graduate program attracted a large number of working professionals, politicians, and potential planning consultants, many of whom were White. Soon after its inception, the program became an important source of planners for the city's planning department and other such agencies, as it was the only program in the region.

The program maintained strong relationships with professional organizations such as the AIP, American Society of Planning Officials, and Association of Collegiate Schools of Planning. Even in these early years graduate students were funded by Housing and Urban Development (HUD) scholarships. Given all these achievements, the program's reputation spread rapidly, and in 1974 it became the first planning program at an HBCU to receive degree recognition—the forerunner of the accreditation process—from AIP. In 1975, the Center for Urban Affairs became the School of Urban Affairs and Human Development and several other degree programs were added.

A Built Environment Studies Program (BES), within the School of Urban Affairs and Human Development, was one of five new urban-oriented programs created under Morgan's urban mandate. The creation of BES led to a change from an M.A. in Urban Planning and Policy Analysis degree to a Master's of City and Regional Planning (M.C.R.P.),[28] the addition of professional graduate degrees in architecture and landscape architecture and an interdisciplinary degree in urban design.

Because of the 1975 legislation that emphasized urban-oriented education, the State Board for Higher Education emphasized that Morgan's three primary missions of teaching, research, and service shall be based on Morgan's dual emphasis on undergraduate and liberal arts programs and *urban-oriented graduate programs*. Morgan will develop into the State's primary public institution dealing with programs that address specific social, political, and economic concerns of *urban areas*. Major program emphasis will include transportation systems, *urban and regional planning and design*, economic development, and programs addressing specific *social problems of cities*.[29] (emphasis added)

Morgan's institutional self-study in the late 1980s further reiterated the relationship between the university's mission and nurturing of planning and related programs. According to this document, the university should be "an integral part of the resource base used by planners, developers and promoters within the Baltimore city, municipal government, the corporate community, and other community organizations of the State."[30] It stated that the university should make every endeavor to inculcate in its student body an understanding of urban America and a sense of social responsibility for improving the quality of life in urban areas. The document further stated that

> Morgan's mission clearly carves out for it a major role in the development and study of Baltimore as a major urban metropolis. It must expand its existing programs and develop new programs that focus on the growing needs of this city, as it typifies urban America. Its current urban-oriented programs in landscape architecture, city and regional planning . . . represent an important reservoir already created by the University to build at Morgan a major instruction and research resource from which the city should draw as it expands.[31]

Facilitating the formation of BES was the underrepresentation of minorities in planning and related professions such as architecture and landscape architecture. In the eyes of the administration, Morgan was able to increase minority representation in these professions because of its historical role in educating Blacks. External factors, such as positive recommendation of a faculty visitation committee, also facilitated funding for BES.[32] The state, on approval of the proposal for BES, appropriated special monies. In addition, in 1978 BES obtained a federal grant under the "Fund to Improve Post Secondary Education." There was ample financial support for the program, in addition to the commitment of the university administration.

The goals of BES were to solve complex contemporary urban issues and to train minority professionals with a high level of technical competence. Initially the planning program took a physical orientation. From the late 1970s to the early 1980s, many of the program's supporters became concerned about this emphasis on physical planning. Furthermore, the program began to lose its reputation, because of the absence of a full-time faculty member as the coordinator, lack of promotion of the program, low profile of the coordinator in Baltimore's communities, and the absence of extension work. In the early 1980s, the arrival of Elva E. Tillman as coordinator rejuvenated the program.

The Planning Accreditation Board (PAB) first accredited the program in 1986. At first the faculty and administration opposed accreditation, fearing that the accreditation process, because of the expense, would exclude minorities and small schools from offering planning programs. At that time the planning program and Morgan were under a financial crisis; faculty had to bear their own expenses to attend conferences, and it was extremely difficult to acquire equipment for the program. In addition, Morgan had only two full-time faculty members, far below the five required for accreditation.

Tillman recalls that prior to the site visit, site visitors assumed that Morgan was resisting accreditation because of the poor quality of the program. However, the team was impressed when they saw the support it had from various professionals, many of whom were graduates of the program. The team was also impressed with the program's community service.

The planning program matured and grew steadily during the 1980s. Enrollment increased because the program was able to fund 75 percent of its students through the HUD work-study program. The administrative home of the planning program changed in 1981 when the School of Urban Studies (formerly School of Urban Affairs and Human Development) merged with the School of Education to form the School of Education and Urban Studies. But Morgan eliminated its urban studies program in the 1980s. The Great Society years were over and there was no need for the quasi-planners that urban studies was producing. Additional problems included the low enrollments, poor quality of students, shortage of faculty, and the inability of other departments (e.g., sociology, political science, and geography) to offer related courses.[33]

Planning, however, survived. The most important event of the early 1990s was the 1991 initiation of the Institute of Architecture and Planning (IAP) as a separate administrative unit to house the architecture, landscape architecture, and planning programs. IAP was created because the accrediting board for university architecture programs required that they be independent administrative units. These changes indicated the university administration's commitment to planning and related programs.

UNIQUE FEATURES OF MORGAN'S PLANNING PROGRAM

Morgan's planning program has certain unique features because of its HBCU home.[34] One such feature is its emphasis on the interests, needs, and concerns of African American planning students. The program willingly reaches out to and admits a moderate number of African American students with somewhat incomplete academic preparation and nonstandard backgrounds but with high motivation and strong promise. Once admitted, however, such students are expected to meet the same standards as other students. Such an approach matches the traditional HBCU method of admitting students who may be underprepared to do collegiate work but have the potential to succeed with proper guidance.

Morgan plays an important role in increasing the representation of African Americans in the profession. A comparison of Morgan's performance with national trends clearly establishes Morgan's role in this area. The percentage of African Americans in accredited masters programs in the United States was 8.9 percent in 1992-1993 and 9.0 percent in 1993-94.[35] Compared to this, Morgan's students were 77 percent African American as of spring 1995. By that time the program had graduated 191 students. Between 1991 and 1995 Morgan graduated twenty-four males and twenty females, of whom thirty-seven, or 84 percent, were African Americans. The program also has sent some of its graduates for doctoral degrees in planning and related fields, fulfilling an important role in increasing minority representation at the doctoral level.[36] Two reasons for the presence of large numbers of African Americans at Morgan are the program's location at an HBCU and its tradition of reaching out to the disadvantaged.

The program focuses on skill development, giving the students, especially the disadvantaged ones, the confidence they need for gainful employment and an opportunity to offer community service.[37] The program follows the traditional advocacy approach to planning education because of the HBCU advocacy role for the disadvantaged.[38] At the same time the program has been a pioneer in dealing with gender, race, ethnicity, and class as an integral part of its curriculum.[39]

Another unique feature of the program is its focus on Baltimore. This is in part dictated by Morgan's urban mission—which carves out for it a major role in the development and study of Baltimore as a major urban metropolis—and its role as an HBCU. As a part of the university's mission of addressing specific social, political, and economic concerns of urban areas, the planning program uses Baltimore as a laboratory for most of its courses. Projects on disadvantaged sections of Baltimore are an integral part of Morgan's curriculum. In addition, the program frequently responds to requests for projects from disadvantaged communities from Baltimore. As an HBCU, Morgan's historic role of reaching out to the disadvantaged further facilitates classroom projects focusing on low-income African American areas of Baltimore. Community projects for Baltimore have been further facilitated by the long tradition of the practitioner faculty, the HUD fellowships, the Institute for Urban Research (IUR),[40] and students who are concerned with community development by virtue of residence, ethnic background, or social concern.

CONCLUSION

Given the context of HBCUs as discussed in this chapter, the existence of planning and related programs at Morgan is almost a paradox. The origins of the programs are attributable to the presence of certain individuals committed to planning or urban affairs, the socioeconomic and political changes in the 1960s, and the institutional changes and culture at Morgan. As an HBCU, Morgan is devoted to educating disadvantaged African Americans. There is an institutional culture of reaching out to underprepared students from inner cities, which in itself is an urban orientation. Morgan's new urban mandate in 1975 and its designation as a public urban university sustains and nourishes planning and related programs. The presence of certain individuals dedicated to planning and the support of the administrators who were willing to interpret the urban mandate as an integral part of planning and related programs created a unique situation at Morgan.

Morgan's goal of admitting underprepared students and nurturing them to be planning professionals provides an important lesson about increasing representation of minorities in the profession. An HBCU can play a special role in increasing such representation by reaching out to the underprivileged, building their confidence, and making them competitive with planners who come from more advantageous backgrounds. Another unique feature of Morgan's planning program—its outreach efforts for disadvantaged African Americans specifically in Baltimore—is also attributable to its HBCU setting.

Unfortunately, HBCU planning programs are becoming a rare entity. Currently, two accredited graduate planning programs remain: the Community Planning and Urban Studies Program at Alabama Agricultural and Mechanical University and the City and Regional Planning Program at Morgan State University. Several challenges are facing HBCU planning programs. For example, Morgan has not received any HUD fellowships since 1993, which has an adverse effect on attracting African Americans. Since HBCUs are not highly endowed or funded, internal fellowships and assistantships that can substitute for external funding (e.g., the HUD fellowships) are few and difficult to obtain. A decrease in fellowships and assistantships can lead to a decrease in enrollment and pose a potential threat to the very existence of programs at HBCUs.

Absence of undergraduate feeder programs (e.g., urban studies/planning, environmental studies/design) exacerbates the enrollment problem. Decreased enrollment makes the program susceptible to the eyes of legislators, especially if there are other such programs in the state. Decrease in enrollment also results in freezing of faculty spots, which may ultimately lead to loss of accreditation and the possibility of the closure of the department. Furthermore, the traditional undergraduate focus of HBCUs makes the sustenance of graduate programs difficult because such programs are often assigned a "stepsister" role. To cite one example, university staff accustomed to dealing with undergraduate students find it difficult to deal with the more mature graduate student demands and needs, including efficient processing of applications. HBCUs still face resource constraints in terms of computers, travel grants for faculty, libraries, and other instructional equipment, despite recent improvements. Lack of such resources can also lead to loss of accreditation and subsequent closure of the program. In summary, HBCU programs face an uphill task to continue their mission of increasing the representation of African Americans in the planning profession.[41]

NOTES

1. See for example, George C. Hemmens, "Thirty Years of Planning Education," *Journal of Planning Education and Research* 7.2 (1988): 85-91; Marshall Feldman, "Perloff Revisited: Reassessing Planning Education in Postmodern Times," *Journal of Planning Education and Research* 13.2 (1994): 89-103. For a good overview of this type of literature, see various issues of *Journal of Planning Education and Research* published since 1981.

2. For exceptions, see, for example, Melvin L. Mitchell, Casey Mann II, and Robert F. Jayson, "The Case for Environmental Planning Education in Black Schools," *Journal of American Institute of Planners* 36.4 (1970): 279-84; Ruth Knack, "Class Acts," *Planning* 53, No. 6 (1987): 28-29; and Elva E. Tillman, "The University's Role in Community Development," in *Planning and Community Equity: A Component of APA's Agenda for America's Communities Program*, ed. American Planning Association (Chicago: Planners Press, 1994), 197-208.

3. See, for example, August Meier and Elliot Rudwick, *Black History and Historical Profession: 1915-1980* (Urbana: University of Illinois Press, 1986); and James A. Banks, "African American Scholarship and Evolution of Multicultural Education," *The Journal of Negro Education* 61.3 (1992): 273-86.

4. This section is based on Julian B. Roebuck and Komanduri S. Murty, *Historically Black Colleges and Universities: Their Place in Higher Education* (Westport, CT: Praeger, 1993) unless otherwise stated. See also Samuel L. Myers, "What is a Black College?" *NAFEO Inroads: The Bimonthly Newsletter of the National Association for Equal Opportunity in Higher Education* 1.5-6 (1987): 1-24; and idem, "What is a Black College?", ibid. 6.5 (1992): 1-5.

5. Roebuck and Murty, *Historically Black Colleges and Universities*, characterize the history of Black higher education into five periods: "antebellum period" (until 1864); "postbellum period" (1865-95); "separate but equal period" (1896-1953); "desegregation period" (1954-75); and "modern period" (1975 to the present). I have not included the "antebellum period" in my characterization, since there were only a few institutions of higher learning in this period.

6. Also see Jane E. Smith Browning and John B. Williams, "History and Goals of Black Institutions of Higher Learning," in *Black Colleges in America: Challenge, Development and Survival*, ed. Charles V. Willie and Ronald R. Edmonds (New York: Teachers College Press, 1978), 68-93.

7. Also see W. Augustus Low and Virgil A. Clift, "Education: Colleges and Universities," in *Encyclopedia of Black America,* ed. W. Augustus Low and Virgil A. Clift (New York: McGraw-Hill, 1981), 338-51.

8. In fact the term *developing institutions* was incorporated to prevent the HBCUs' being the only institutions that were eligible for federal funding. *Adams v. Richardson,* 36 F. Supp. 92.

9. Also see Garry Boulard and B. Denise Hawkins, "Fordice Judge Visits Black Campuses in Mississippi," *Black Issues in Higher Education* 11.22 (1994): 6-9.

10. In addition to the references, this section is based on informal discussions (August 1992 to June 1995) and a formal interview (May 1995) with Anthony N. Johns Jr., director of the Institute of Architecture and Planning at Morgan (IAP); an interview and an informal telephone conversation (June 1995) with Elva. E. Tillman, an ex-coordinator of the City and Regional Planning Program at Morgan; informal conversations (August 1993 to June 1995) with George Worthy, associate professor of city and regional planning at Morgan.

11. Roebuck and Murty, *Historically Black Colleges and Universities.*

12. Garland L. Thomson, "Engineering, Business Majors Grow Among Blacks," *Black Issues in Higher Education* 11.22 (1994): 10-11.

13. Mitchell et al., "The Case for Environmental Planning Education in Black Schools."

14. Ibid.

15. Cheryl K. Contant, Peter S. Fisher, and Jennifer R. Kragt, *Guide to Graduate Education in Urban and Regional Planning,* 9th ed. (Chicago: American Planning Association, 1994).

16. Anthony Johns, interview. Johns was associated with Howard's architecture department from 1963 to 1985. Mitchell, et al., "The Case for Environmental Planning Education in Black Schools."

17. For an overview of these programs, see, for example, Susan S. Fainstein and Norman I. Fainstein, "Economic Change, National Policy, and the System of Cities," in *Restructuring the City: The Political Economy of Urban Redevelopment,* ed. Susan S. Fainstein, Norman I. Fainstein, Richard C. Hill, Dennis R. Judd, and Michael P. Smith (New York: Longman, 1983), 1-26.

18. Lewis R. Fibel, "Education for Urban Assistants," *Planning 1966* (Chicago: American Society of Planning Officials, 1966), 166-68; and Franz J. Vidor, "A Junior College Program to Train Planning Assistants," *Planning 1967* (Chicago: American Society of Planning Officials), 132-39. Note that community colleges without accredited programs are still sources of employees for local professional needs. See, for example, Thomas D. Galloway, "Threatened Schools, Imperiled Practice: A Case for Collaboration," *Journal of American Planning Association* 58.2 (1992): 229-34.

19. See Roebuck and Murty, *Historically Black Colleges and Universities.*

20. Mitchell et al., "The Case for Environmental Planning Education in Black Schools." The Moton Consortium on Admissions and Financial Aid, *The Moton Guide to American Colleges with a Black Heritage* (Atlanta, GA: Stein Printing Company, 1973); idem, *The New Moton Guide to American Colleges with a Black Heritage* (Richmond, VA: William Byrd Press, 1983).

21. Moton Consortium, *Moton Guide;* idem, *New Moton Guide.*

22. This section is based on Morgan State University, *Morgan State University Catalog, 1984-1986* (Baltimore, MD: Morgan State University, 1986); idem, *Institutional Self-Study: Prepared for Middle States Association of Schools and Colleges* (Baltimore, MD: Morgan State University, 1988); and various other reports of the university.

23. Due to space constraints the author cannot provide the comprehensive list of interviews and reports that are the basis for this section of the chapter. Readers interested in the complete citation may contact: Siddhartha Sen, Institute of Architecture and Planning, Morgan State University, Jenkins 334, Baltimore, MD 21239-4098. See also the list of interviewees and reports in note 34.

24. There were a few urban related courses and programs prior to the early 1960s. These were generally efforts pioneered by individual faculty members and hence fall outside the "formal institutionalization" of planning and related programs.

25. For a good discussion on the topic, see Matt Gottdiener, ed., *Cites in Stress: A New Look at the Urban Crisis*, vol. 30 (Beverly Hills, CA: Sage, 1986).

26. In this context, see Howard E. Mitchell, ed., *The University and the Urban Crisis* (New York: Behavioral Publications, 1974); and Thomas P. Murphy, ed., *Universities in the Urban Crisis* (New York: Dunellen Publishing, 1975).

27. Morgan State College, *Proposed Center for Urban Affairs* (Baltimore, MD: Morgan State College, 1970), iv.

28. The change of name from M.A. in Urban Planning and Policy Analysis degree to M.C.R.P. was initiated because the M.C.R.P. degree was popular and a common name for graduate degrees in planning at that time.

29. The State Board for Higher Education as cited in *Morgan State University Catalog, 1984-1986*, 13.

30. Morgan State University, *Institutional Self-Study: Prepared for Middle States Association of Schools and Colleges*, 8.

31. Ibid., 9-10.

32. The proposal for BES was submitted to the State Board for Higher Education in December 1977 and was approved in October 1978. BES was made operational in the 1979-80 academic year, when it admitted first groups of students in architecture and landscape architecture.

33. Anthony Johns, interview.

34. This section is based on interview and telephone discussion with Elva E. Tillman; informal discussions and an interview with Anthony N. Johns Jr.; informal discussions with George Worthy; informal discussions with Mahendra H. Parekh; interview and informal discussions with Shirley R. Byron; Institute for Urban Research (IUR), *Institute for Urban Research at Morgan State University* (Baltimore, MD: IUR, Morgan State University, n.d.); idem, *Continuing the Response to an Urban Mission: The Institute for Urban Research* (Baltimore, MD: IUR, Morgan State University, 1995); and self-study reports for initial accreditation and reaccreditation of the City and Regional Planning Program.

35. See Contant, Fisher, and Kragt, *Guide to Graduate Education in Urban and Regional Planning*.

36. In this context, also see Edward W. Hill, "Increasing Minority Representation in the Planning Professoriate," *Journal of Planning Education and Research* 9.2 (1990): 139-42.

37. In this way we are attempting to teach what has been described as "savvy." See Karen S. Christensen, "Teaching Savvy," *Journal of Planning Education and Research* 12.3 (1993): 202-12.

38. For a detailed discussion of advocacy planning, see, for example, Paul Davidoff, "Advocacy and Pluralism in Planning," *Journal of American Institute of Planners* 31.4 (1965): 331-38; Paul and Linda Davidoff and Neil N. Gold, "Suburban Action: Advocate Planning for an Open Society," *Journal of American Institute of Planners*, 36.1 (1970): 12-21.

39. Note that these issues are being debated as tasks for planning education in the late twentieth century. Among others see for example, John Friedmann and Carol

Kuester, "Planning Education for the Late Twentieth Century: An Initial Enquiry," *Journal of Planning Education and Research* 14, no. 1 (1994): 55-64.

40. The IUR was established under provision of the Maryland State Legislature as a component of the School of Graduate Studies in 1978. From its inception IUR was independent of BES, although it maintained strong relationships with it. IUR was also a response to the university's urban mission and carries out research, training, and community service.

41. I would like to thank Professor June M. Thomas of Michigan State University; Professor Marsha Ritzdorf of Virginia Polytechnic and State University; Elva E. Tillman, assistant city solicitor, City of Baltimore; and Professor Emeritus Thomas A. Reiner of the University of Pennsylvania for their constructive criticisms on earlier drafts. I would also like to thank my colleagues at Morgan State who provided a wealth of information through interviews and innumerable conversations.

15

COMING TOGETHER

Unified Diversity for Social Action

JUNE MANNING THOMAS

Insights into the relationship between U.S. urban planning and the African American community do little good if not put into practice.[1] The main purpose of improving knowledge about the relationship between race and planning should be to influence planning practice for the better, and the best place to begin to improve such knowledge is the planning classroom.

Of course, race is not the only social issue needing attention in the classroom. Educators should include more information about gender, class, and international thought in planning courses. An appropriate model for planning education would train effective planners and would also champion diversity and positive action for social reform. One possible model, "unified diversify for social action," would allow plan-

AUTHOR'S NOTE: This chapter is a condensed version of an article "Coming Together: Unified Diversity for Social Action," *The Journal of Planning Education and Research* 5 (3), and is used here by permission.

ning schools to foster the education of all students; would reflect different perspectives of race, gender, and national origin; and would simultaneously improve the ability of future planners to function well in the contemporary urban environment.

These tasks, which may seem at first hopelessly complex, are extremely important. The workforce is changing dramatically, and societal inequities linger. *Multiculturalism*—a term that typically includes race, ethnicity, gender, and nationality—has emerged in planning education, but with little clarity, cohesion, or purpose. Embracing a stronger, more activist vision for diversity may create a new focus for fragmented, conflict-ridden efforts and help train more effective planners than can traditional educational programs. This chapter explains how this is possible, by focusing particularly on the role of race—and, to a lesser extent, gender and internationalism—in planning education.

PLANNING DIVERSITIES?

A new era of "pluralism" or multiculturalism has already dawned. When two researchers recently asked thirty-two planning faculty members what were the greatest national and local challenges facing planning education in North America, respondents indicated that an important planning skill was the ability to plan for multicultural diversity in the metropolis.[2] Yet the process of teaching planners how to do this has been uneven. Academic programs are just beginning to acknowledge the contributions of women to the urban environment or the practical implications of their concerns for practicing planners. Minority representation among faculty and students has stagnated, and little cohesion informs the treatment of race within the planning curriculum.

Internationally oriented students and faculty have rightfully challenged the cultural biases of U.S. planning education. They have called attention to the impact of different languages and thinking styles, and they have questioned the tendency to compel students to learn planning "the Western way," as if all knowledge stemmed from one country and one culture.[3] They have suggested the alternative of "mutual learning," whereby domestic and international students share their experiences in a reciprocal manner and focus on global linkages and common concerns.[4]

The dialogue concerning diversity issues related to planning education for racial minorities and women is growing but, by comparison, is less well developed, a fact that one can discern simply by reviewing back issues of the *Journal of Planning Education and Research*.[5] Yet race and gender are important concerns for U.S. students, the base for U.S. planning programs.

The number of women planners is growing markedly. In 1968, 7.5 percent of U.S. planning program graduates were women; by 1978, that proportion had increased to 31 percent, and by 1994, 45.4 percent of planning masters' students were female.[6] Surveys of the American Planning Association (APA) showed that female membership rose from 18.2 percent in 1981 to 27.7 percent in 1991. Yet the upper ranks in the profession were largely male.[7] Furthermore, in 1994, male planning faculty (teaching planning 50 percent or more) outweighed female faculty five to one.[8] Little wonder that many students graduate from planning programs with little or no information about women's contributions to the profession or about the relationships between gender, poverty, housing, and land use.[9]

The record on race is mixed. One recent survey showed that, in North American planning schools, African Americans made up 12.6 percent of master's students, comparable to their overall proportion in the population, but they made up only 6.4 percent of doctoral students. Another 1994 survey found that 9 percent of U.S. planning master's students were Black, with lesser percentages for other minorities.[10] As for African American faculty, numbers barely crept up from 1968-76, when fourteen received full-time faculty appointments, to 1994, with thirty-four total, making up only 5.7 percent of all planning faculty.[11] The numbers are even smaller for Asians, Hispanics, and Native Americans.

An additional concern is the way the planning literature and curriculum have dealt with matters of race. Potential minority planning students may weather survey or introductory planning courses that give them precious little reason to become interested in planning. Planning history may give the impression that Blacks have had no role in improving cities, or a role only as victims, and may mention other minorities not at all. Courses that stress land use planning and technical specialties may seem irrelevant at best, oppressive at worst, if they reinforce the existing social order of inequity and separatism.[12]

This state of affairs is amazing considering how much racial disunity affects the work of urban planners. Scarcely a U.S. rural area or suburb is so remote that it does not experience the social pressures of racial segregation. In metropolitan areas, historic and contemporary racial estrangement severely hinder urban improvement. In large part, the true "urban problem" facing today's metropolitan areas is old-fashioned racial and income segregation.

As Goldsmith pointed out in a critique of an Association of Collegiate Schools of Planning (ACSP) report on education, it is too easy to forget about the importance of racial and ethnic diversity.[13] ACSP has sponsored a series of reports exploring diversity in planning education, which have led to new accreditation standards for planning schools.[14] These will undoubtedly prove helpful. But planning educators must also embrace diversity for other reasons besides accreditation requirements. Students must learn to work together on the job, since the planning office is not exempt from discrimination; surveys show that minority and women planners experience persistent problems with salary disparities and lack of access to informal networks.[15] Future planners must understand the importance of opposing discrimination and upholding the rights of all urban residents. They must also be able to help bring about effective change in the contemporary metropolis.

THE NONPLANNING
LITERATURE ON DIVERSITY

The "diversity" literature suggests possible ways to carry out the balancing act between improving diversity and insuring effective training. Relevant guidance comes from both the organizational literature, which focuses on the workplace, and the educational literature, which addresses the classroom.

The literature on multicultural organizations is large and growing.[16] This literature, based largely on corporate experience, clarifies the positive results of diversity in the workplace and suggests stages of growth in that process. Roosevelt Thomas identifies three such stages: affirmative action, "valuing differences," and "managing diversity." The last stage modifies the core culture so that it works for everyone and increases workplace productivity.[17] Cox, who promotes the concept of the "multicultural organization," also proposes three stages: monolithic organizations, which are homogeneous and expect assimilation by any minority; plural organizations, which are more heterogeneous but still rely on assimilation; and multicultural organizations, characterized by diverse workers, inclusion of minorities in networks and social activities, absence of prejudice and discrimination, and low levels of intergroup conflict.[18]

The multicultural education field, which is at least thirty years old, has already affected elementary, secondary, and university schooling in countries around the world.[19] As this literature notes, knowledge is not really "objective" or "neutral"; instead, it reflects the assumptions, biases, and culture of those who create it. When the creators are in positions of power, their vision of truth prevails. For example, a Eurocentric view of history will result if the only people writing history come from Eurocentric backgrounds and training. Multicultural education expands cultural vision, showing familiar issues from different perspectives and granting credibility to cultures and perspectives often considered inferior or unimportant.[20]

Educational institutions tend to favor society's core culture. In North America, that culture values individual social mobility rather than group loyalty, and it rewards linear thinking, the kind of mentality that makes bureaucracies flourish. Although it is dangerous to typecast individuals, members of some microcultures do not function best according to these values. In certain Latino, Native American, and African American populations, the group is the locus of reality, so learning environments oriented to individuals pose significant barriers.[21]

Multicultural education, properly approached, can make learning easier for different races, nationalities, and racial or ethnic minorities. It can enable students of all backgrounds to develop more positive attitudes toward people who are different. It can "empower students from victimized groups" to "develop confidence in their ability to succeed academically," as well as help them "influence social, political, and economic institutions."[22] The multicultural education literature makes concrete suggestions for fostering such education. Some teachers have developed ingenious ways to teach undergraduates to value their own diversity—even if they come from supposedly homogeneous "White" backgrounds—and thus become more accepting of differences.[23] Others have recommended institutional and administrative changes or mentoring programs to ensure retention of women and people of color.[24]

As with the literature about organizations, multicultural education suggests phases of change. Tetreault offers five phases for including gender issues, ranging from (1) a strictly male-defined curriculum to one that (2) acknowledges the contributions of women, (3) presents women's concerns as dichotomized from males, (4) uses women's activities as the norm, and finally, (5) balances concerns of women and of men. Banks suggests expanding racial and ethnic issues via four phases. The first inserts ethnic heroes and artifacts into the curriculum; the second adds content and concepts about race, although in a poorly integrated fashion; the third views different themes and problems from different ethnic perspectives; and the last encourages students to become "reflective social critics" dedicated to social change.[25]

PLANNING EDUCATION MUST EVOLVE

The challenge to planning education is to grow more consciously beyond mainstream-centered perspectives, in a way that does not destructively fragment the whole (hence, the need for *unified* diversity) but, rather, supports effective planning and social reform. Adopting the simple technique of viewing diversity in stages helps clarify where planning education has come from and where it needs to go. At least three phases of growth are apparent: monoculture, pluralism, and unified diversity. Table 15.1 summarizes how these phases compare with three other phased educational models, and Table 15.2 summarizes characteristics of each.

Monocultural Planning Education

Although monocultural elements survive in the present, this phase flourished from the beginning of U.S. planning education until the mid-1960s. Planning reflected modernist thought and logical positivism: it was technical and supposedly value-neutral, governed by rational rules. International students were obliged to accept U.S.-based knowledge so as to help their countries "develop." Planning schools trained students in land use planning without particular reference to race or gender. Faculty and student body were largely White and male, as were the authors of their texts.

Cities were certainly diverse, as was academia, to a limited extent. Key women scholars in the 1930s and 1940s included housing expert Edith

Table 15.1 Selected Models of Educational Multiculturalism

	Tetreault 1989	*Banks 1989*	*Sanyal 1989*	*Thomas 1996*
Focus	*Gender*	*Race and Ethnicity*	*Internationalism*	*Race and Gender*
Phases	1. Male-defined	1. Contributions	1. Western dominance	1. Monoculture
	2. Contribution	2. Additive	2. Core/ periphery, "alternative" theories	2. Pluralism (race, gender, international)
	3. Bifocal	3. Transformation	3. Global education	3. Unified diversity for social action
	4. Women's	4. Social action		
	5. Gender-balanced			
Primary educational system	K–12	K–12, university	Planning education	Planning education

SOURCES: Mary Tetreault, "Integrating Content About Women and Gender into the Curriculum," in *Multicultural Education: Issues and Perspectives,* ed. James Banks and Cherry Banks (Needham Heights, MA: Allyn and Bacon, 1989), 124-40; James Banks, "Integrating the Curriculum with Ethnic Content: Approaches and Guidelines," in ibid., 192; Bishwapriya Sanyal, "Poor Countries' Students in Rich Countries' Universities," *Journal of Planning Education and Research* 8 (1989): 139-56.
NOTE: Phases are inferred from text of Sanyal. Ideas in last column developed by present author.

Elmer Wood and redevelopment thinker Catherine Bauer Wurster, and material feminists were already reshaping concepts of home and community.[26] One could also consult several distinguished African American scholars on the subject of urbanism, such as St. Clair Drake and Horace Cayton or W. E. B. DuBois. A few scholars, such as Wurster, wrote about the interaction of race and planning activity.[27] But the planning academy was, on the whole, monocultural. Few women and racial minorities were enrolled in planning schools, because of accepted practice or explicit bias against entrance.[28]

Pluralist Planning Education

For at least thirty years, the pluralist era has prevailed. It arrived in the 1960s, when Perloff noted an "exponential growth" in the number of minority students, when writers such as Altshuler began to question the

Table 15.2 Paradigms of Planning Education

	Monoculture	*Pluralism*	*Unified Diversity*
Program administration/ Goals	Technical	Eclectic, disjointed	Affirmative, reformist
Faculty			
Membership	White, male	Interim levels of diversity by gender, ethnicity, and race	Diverse, cooperative, supportive
Style	Hierarchical	Eclectic, contradictory	Interactive, dialogue-based
Student body	White male	Moderate diversity, segmented	Extensive diversity, cooperative
Academic environment	Not supportive of "differences"	More inclusionary, but incomplete support for those "different," disjointed	Inclusionary, whole group identity, mentoring, and support
Curriculum	Ethnocentric, rationalistic, monocultural	Contributional, bifocal, or additive; poorly integrated; occasional inclusion of diversity	Diverse, transformational, well-integrated, good preparation for effective social action

rational paradigm, and when the urban riots and advocacy planning challenged the old models of value-neutral planning.[29]

During the pluralist phase, planning education experienced at least three waves of change related to diversity. The apparent "exponential" growth of minorities and minority issues in the 1960s did not last long. The 1970s and early 1980s produced vacillating concern for social, advocacy, and equity planning, all tied in some way to racial justice but not quite the same thing. Then the feminist revolution brought a new wave of writing and research, which is still growing and evolving. Dialogue about planning education for internationalism also emerged. But advocates for these various movements remained remarkably disconnected.

As the literature on racial conditions in U.S. metropolitan areas exploded, a dizzying array of urban studies issued forth concerning the interaction of race and employment, housing, social welfare, redevelopment, and other issues of critical importance. Several publications focused on planning and redevelopment. The heightened sensitivity to race pushed practitioners and scholars to consider issues of justice and equity and forced the emergence of social, advocacy, and equity planning. By the early 1990s, however, the volume of literature on issues of planning and race appeared to diminish.

In the meantime, production of materials on gender and planning increased. At first much of this was in the contributional mode identified by Tetreault and Banks, as it "rescued" women from the abandoned heaps of history and made their contributions and visions known. But the literature soon gravitated to a more critical phase of multiculturalism, moving toward bifocalism, as with those authors who focused on biases against women in urban space or in planning actions. Some recent writings resemble Tetreault's "women's curriculum" stage, as does Hendler's call for "the substitution of feminist for traditional ethics" in planning education, an astonishing statement that would have been unimaginable a few years ago.[30]

Few planning faculty seemed persistently interested in the relationship between race and planning education, however. For a few years the Faculty Women's Interest Group sponsored miniseminars to consider both race and gender in the curriculum. For many years, participation in seminars about internationalism has included relatively few women or indigenous U.S. minorities.[31] Part of the weakness in dialogue about the relationship between race and planning education relates to the stagnation in numbers of minority planning faculty and to the decline in the number of planning programs at historically Black institutions (see Sen, Chapter 14, this volume).

The clashing of the old, traditional cultures and the new ones is also of concern. For example, African American faculty and students may see things differently than do majority-race colleagues. Expecting minority students to sit passively in a planning classroom and study nothing about themselves and their life experiences may lead to discontent and uneven performance. Treating ethics as a matter of professional procedure, without instilling social values, makes the purpose and intent of planning unclear. Field projects that focus on the comfort of U.S. society's mainstream (middle-class White suburbia) are often unexciting and send a strong message of conformity.

Students may become restless. Programs that went to great extremes to recruit minority students may lose them to monocultural course content. Outnumbered minority (or women) faculty often struggle to lead recruitment, mentor those students who might leave, and carry out their own research. These faculty may leave themselves, particularly if monocultural faculty networks shut them out. They may hear constant comments from White male students and faculty about how "different" their courses are than those of other faculty, in style as well as content, making the program appear disjointed. The message is that their courses don't "fit" and that their contributions are less important than those of the mainstream.

These are symptoms of disjointed pluralism, incompletely digested diversity. A planning program with these problems, whether hidden or readily apparent, stands at a stage that multicultural experts suggest making the effort to move through.

Stronger planning education accreditation requirements will doubtless lead to further pressure to promote diversity in planning education. These standards encourage recruitment of students and faculty and urge programs to address racial, ethnic, income, and gender topics in courses.[32] Standards hardly tell schools how to do this, however. Requiring affirmative action is merely a first step and could generate only half-hearted compliance, especially as the changes in federal and state affirmative action compliance leave the accreditation requirements without substantive backup at higher institutional levels.

Unified Diversity for Social Action

We need to identify a third stage, one that goes beyond monoculturalism and disjointed pluralism. This stage should be comparable in advancement of evolution to Roosevelt Thomas's third and highest stage, "managing diversity"; Cox's third stage, the "multicultural organization"; Tetreault's fifth stage, a "gender-balanced curriculum"; Banks's fourth stage, "social action approach"; and Sanyal's (implied) third stage, "one world" planning education.

This is a dizzying array of terms and options. What all of these higher stages have in common, first of all, is that they are visionary, describing desired states of being that may not yet exist completely. That is, good examples may be hard to find, but the point is to visualize them and aim toward creating them. In addition, each of these higher stages implies tolerance and cooperation but also suggests a higher level of organizational or educational quality and effectiveness than previous levels.

Banks chose the term *social action* to characterize his highest stage because he realized that the most powerful positive form of race and ethnic studies enabled students to "make decisions on important social issues and take actions to help solve them."[33] The phrase "unified diversity for social action" fits a culminating phase for multicultural planning education because it not only acknowledges the role of diversity but also identifies its major purpose. That purpose, as stated by Perloff, is to respond to the question, "How can we train the planners we need in our efforts to achieve a free, fruitful and peaceful life?"[34] This connects with the historical social reform agenda of planning.

Based on the multicultural literature reviewed thus far, the following would seem to be important characteristics of urban planning academic programs exemplifying unified diversity for social action (also summarized in Table 15.2):

- *Administration:* Leadership in, and support for, effective and unified diversity for social action; active planning for improved program performance in recruitment, retention, and other areas of concern, simultaneously encouraging high standards of education, scholarship, and public service; leadership in the concept of social action for reform in the urban and regional environment
- *Faculty:* Diverse membership, of race, ethnicity, gender, and nationality, according to the limitations of program size and recruiting flexibility; meaningful integration of pedagogical styles, with sharing and interactive improvement of techniques and approaches based on multicultural perspectives, all in the context of a curriculum that is balanced and complementary; cooperation and comradeship among faculty, with no exclusion based on race, gender, or other exterior characteristics
- *Students:* Again, diversity of race, ethnicity, gender, nationality, and class background, according to the limitations of size and resources; admission according to typical criteria for educational achievement but with dedication to improving or maintaining diversity, seeking in particular those with experience with—or potential dedication to—effective social action in the planning field
- *Academic environment:* Support for a variety of cultures, backgrounds, and learning styles, along the ranges of individualistic/group, rationalistic/holistic, and so forth; inclusion of the cultural experiences of students in course and noncourse activities; encouragement of group identity

among planning students as a whole as opposed to oppositional or exclusionary cliques based on race, gender, class, or other such characteristics; active mentoring between faculty and students, and "older" students and "younger" students

- *Curriculum:* A planning curriculum that reflects the best of multicultural knowledge about urban society and planning practice and that provides for meaningful social action in the context of superior skill enhancement, possibly in practicum or field studies courses or in regular course components

Further Thoughts on Curriculum

It is not possible, within the scope of this chapter, to offer a complete overview of curriculum areas related to multicultural planning education.[35] Programs and instructors need to tailor topics and materials to fit their classes and in some cases may prefer to provide separate classes on race or gender. Eventually, however, planning programs should integrate diversity issues into the regular curriculum, to ensure wide exposure. The following areas appear to offer particular opportunities for such integration, in the context of recently revised accreditation requirements for planning schools.

Urban Society. Accreditation requirements state that planning students should gain knowledge about the city and its regional context, for a range of subject areas, "including multicultural and gender dimensions."[36] Introductory courses on urbanization provide an excellent opportunity to draw on the extensive literature concerning race, ethnicity, and gender in urban society; some of that literature is directly planning related.

Planning History and Theory. Accreditation requires planning schools to teach the history and theory of planning "in relation to . . . such characteristics as income, race, ethnicity, and gender."[37] Planning history offers an exceptional opportunity to include more subject material related to gender, race, and ethnicity. One task is to move beyond examining "contributions" and existing conditions and to instead provide "transformational" course material, as with the third stage of Banks's model. This means helping students learn how to view issues from the perspective of diverse cultural, racial, and ethnic groups.

Planning theory is variably defined—recent surveys suggest little consistency in course materials—but can include skill-building topics such as mediation and negotiation or strategic planning. Such topics are of particular value for social action, with the addition of the somewhat less popular topics of citizen participation and social change.[38] Special attention to planning reform traditions such as advocacy and equity planning offers the opportunity to discuss issues of social inequity and diversity, particularly diversity of race. Connections between planning theory and feminist theory are well documented and evolving as a field of study.[39]

Ethics. This area, included by some programs in planning theory courses, deserves special attention. A fundamental requirement is for students to study and absorb those provisions of the professional American Institute of Certified Planners (AICP) code that relate to racial minorities, women, and the disadvantaged (see planning-related documents, Part 5, this volume).[40] Understanding the contributions of feminist ethics will be helpful, as will exploration of the ethical implications of society's racism.[41] It is even more important to help students understand why planners should assist in improving society. To do so, we must recapture the normative tradition in planning thought and "develop a greater sense of moral obligation and social responsibility."[42]

Law/Land Use. Race and gender-related topics abound for planning law or land use classes. The history of zoning laws regarding exclusion by race, income, or family status offers important background material for all planning students.[43] Exploring modern cases of exclusion and court-ordered acceptance of subsidized housing, such as in Yonkers, New York, and Mt. Laurel, New Jersey, can help students prepare for the pressures that will face them in suburban or regional planning.[44] Classes might also examine the relationship between metropolitan land use and the status of women and families,[45] or the isolation of low-income people and racial minorities.[46]

Methods. Planners still use the rational paradigm, so quantitative methods of analysis are important. Davidoff raised key points about the possibilities of addressing issues of racial inequity in traditional

analysis classes. He suggested that "distributional analysis of the allocation of resources, wealth, income, opportunity and knowledge in
society should be a central component of planning analysis education."[47] It may also be important to focus more consciously on qualitative methods of inquiry, however. Dalton urges planners not to judge
qualitative approaches as irrational but, rather, to respect other modes,
such as "experience coupled with intuition, leadership, and bril-
liance."[48] Her support of humanitarian or empathetic understanding as
valid is a point supported by feminist scholarship.[49]

Specialty Areas. Accreditation requirements indicate that a planning
student should study at least one specialty area, which may include
"housing, land use, economic development, urban design, the environment, and transportation," among others. "Wherever appropriate,"
however, areas should consider "racial, ethnic, income, and gender-
related issues."[50] Although discussing the endless alternatives is impossible here, materials concerning race, ethnicity, and gender are available
for each of the preceding areas.

Field Study, Community Service Course Modules, and Guided Internships.
Fieldwork experience is important because it simplifies the complexity
implicit in the preceding dialogue by applying theory to practice. It is
in the context of experience that theory comes to life; students can see
for themselves the results of racial and income inequality and grapple
with using planning tools to help address this issue. Experiential learning, which "involves the whole person in thinking, feeling, and acting,"
has many advantages over "learning acquired by listening to lectures,
reading books, and analyzing statistics."[51] And so one of the most
effective techniques for teaching effective social action is to involve
students in a guided experience that does just that.

A simple way to help planning students understand how to plan in a
diversified world is to assign them to work with community groups in
low-income mixed-race or minority neighborhoods. In this context,
students learn more deeply about planning concepts such as neighborhood, district, and master planning, but they also learn about other areas
such as community development, community organizing, and city governance, and they simultaneously gain exposure to working with other
races and income groups.

A personal example may help to illustrate the potential benefits of field work for such instructional purposes. When the author first began to develop fieldwork courses in inner-city Detroit, racial divisions posed difficult challenges. One year, all the students were White (except for one Asian), and all the residents who retained the students' services were Black. The students did their best to overcome culture shock (as did the residents). They learned much about common humanity and also gained creditable neighborhood planning skills. Possibilities for using planning to improve central cities opened up to them, as they saw the dedication of neighborhood residents and as they worked to assist neighborhood planning efforts. Yet the students' basic skill development did not suffer; in fact, planning in such a low-income neighborhood forced them to refine these skills more carefully than would have been the case otherwise.

Even though the project turned out to be a good experience for all involved, the differences between that year and subsequent years when the students included African Americans were palpable. The level of comfort increased, and the positive interactive effects of several races of students, as well as both genders, seemed to heighten dedication and effectiveness. Minority community placement was able to provide a more "diversified" experience for planning students no matter what their color, but this effect was heightened if the students themselves were diverse.

THE VISION

Planning educators have only begun to explore the issue of multiculturalism from the perspective of urban planning. On the whole, planning education has left behind a strictly monocultural phase, and it has reached various phases of a rather disjointed pluralism. Yet the absence of unity, common vision, and perhaps enthusiasm for the topic of diversity leads one to believe that planning education needs to move to a third stage, here labeled "unified diversity for social action," and comparable to the "higher" stages that researchers in multicultural corporations and educators have called most effective and beneficial.

What would this look like? Surely planning schools characterized by unified diversity would retain extensive but harmonious diversity in the faculty, student body, and curriculum. In such planning programs, people would celebrate diversity as a source of strength and creativity. Picking up on Banks's educational ideas, monoculturalism would pose no barriers to learning, and course material would consider multiple perspectives on urban planning history, theory, and practice. The learning environment would be supportive of women and racial minorities. The focus on a common end—social justice and improvement for residents of urban and regional communities—would yield unity of purpose.

In a planning program characterized by unified diversity, one would expect to see varied backgrounds among students and faculty. Students would benefit from proper preparation for work in the contemporary metropolis, having received strong overt and covert messages that all races and cultures are valuable, as are both sexes, and that diversity is a strength to be nurtured rather than a requirement to be tolerated. As with Cox's multicultural organization, minorities and females would be included in networks and social activities, absence of prejudice would be the norm, and little intergroup conflict would exist. The curriculum would provide the best possible preparation for students to plan effective strategies for helping create a better urban and regional society.

A more diversified profession can offer new insights and creativity concerning critical problems facing the North American metropolis, such as the fragmentation of the region, the decline of the central city, and the extremes of wealth and poverty. Having more planners of African American and other minority racial backgrounds, who belong to several cultures and understand the context of many city neighborhoods, can enhance trust between residents and planners and perhaps lay the groundwork for more creative strategies for city improvement. Such professionals may also gain greater connection with the minority mayors now in office. Bringing minority youth into the profession could also assist upward mobility and professional growth among the oppressed races, a task shouldered to a disproportionate degree by the dwindling number of planning programs at historically Black universities.

In such ways, embracing a stronger vision for diversity could train more effective planners for tomorrow. Society has changed and so must those who work in it. Planners who most consciously embrace the positive aspects and benefits of such changes are best prepared to respond effectively to the contemporary metropolis. It will be an enormous challenge to implement these changes, but it is a task that planning educators can successfully address.

NOTES

1. Many of the insights gained in this chapter stem from the author's experiences, which include twenty years of experience as the only African American faculty member in a planning school, membership on the Diversity Committee of the Association of Collegiate Schools of Planning, and co-ownership of a private consultant firm focusing on improving race relations in corporations and schools. Membership in the Bahá'í Faith, a worldwide religious community devoted to the unity of humankind and the elimination of prejudice of all kinds, gave the author important practical experience in promoting diversity and understanding its benefits. The chapter is a revised version of a spring 1996 article published in the *Journal of Planning Education and Research* (reprinted by permission).

2. John Friedmann and Carol Kuester, "Planning Education for the Late Twentieth Century: An Initial Inquiry," *Journal of Planning Education and Research* 14 (1994): 55-64.

3. Jay Chatterjee, "Why New Perspectives are Needed," in *Breaking the Boundaries: A One-World Approach to Planning Education*, ed. Bishwapriya Sanyal (New York: Plenum Press, 1990); Louise Dunlap, "Language and Power: Teaching Writing to Third World Graduate Students," in ibid.

4. Bishwapriya Sanyal, "Poor Countries' Students in Rich Countries' Universities: Possibilities of Planning Education for the Twenty-first Century," *Journal of Planning Education and Research* 8 (1989): 139-56.

5. The *Journal of Planning Education and Research* (published since 1981) has published far more articles on planning education issues related to internationalism than on planning education related to race or gender.

6. Jackie Leavitt, "The Gender Gap: Making Planning Education Relevant," *Journal of Planning Education and Research* 3 (1983): 55-56. The 1994 data are from Cheryl K. Contant, Peter S. Fisher, and Jennifer R. Kragt, *Guide to Graduate Education in Urban and Regional Planning*, 9th ed. (Chicago: American Planning Association, 1994), xv.

7. Judith Getzel and Gregory Longhini, *Planners' Salaries, Reports, and Trends, 1981*, Planners Advisory Service Report no. 366 (Chicago: American Planning Association, 1982); Marya Morris, *Planners' Salaries and Employment Trends, 1991*, Planner Advisory Service Report no. 439 (Chicago: American Planning Association, 1992).

8. Contant et al., *Guide to Graduate Education*, xiv.

9. Leonie Sandercock and Ann Forsyth, "A Gender Agenda: New Directions for Planning Theory," *Journal of the American Planning Association* 58 (1992): 49-59; Edith Netter and Ruth Price, "Zoning and the Nouveau Poor," *Journal of the American Planning Association* 49 (1983): 171-81.

10. James H. Mars and Joe Springer, "Who Are We Serving? A Comparison of the Students in Graduate and Undergraduate Programs," paper presented to the Association of Collegiate Schools of Planning, Philadelphia, October 1993. Contant et al., *Guide to Graduate Education*, xv, which focused on 1994 U.S. programs only, found 9.0 percent of

planning students were Black, with 8.6 percent Hispanic, and 11.2 percent of non-U.S. origin.

11. Catherine Ross, "Increasing Minority and Female Representation in the Profession: A Call for Diversity," *Journal of Planning Education and Research* 9 (1990): 135-38. The 1994 data are from Contant et al., *Guide to Graduate Education*, xiv.

12. Robert Catlin, "The Planning Profession and Blacks in the United States: A Content Analysis of Academic and Professional Literature," *Journal of Planning Education and Research* 13 (1993): 26-32; June Thomas, "Planning History and the Black Urban Experience," *Journal of Planning Education and Research* 14 (1994): 1-11.

13. Association of Collegiate Schools of Planning, Commission on Undergraduate Education, "Creating the Future for Undergraduate Education in Planning," *Journal of Planning Education and Research* 10 (1990): 15-26; William Goldsmith, "Remarks on the Niebanck Commission Report on Undergraduate Education," *Journal of Planning Education and Research* 11 (1991): 75-77.

14. Committee on the Recruitment and Retention of Women and Minorities in Planning Education, "The Recruitment and Retention of Women Faculty and Faculty of Color in Planning Education: Survey Results" (Madison, Wisconsin: Association of Collegiate Schools of Planning, 1990); J. Eugene Grigsby III, "Minority Education in Planning Schools" (Unpublished report, 1988); Planning Accreditation Board (PAB), "Accreditation Criteria Changes in *The Accreditation Document*," handout based on changes approved at May 7-8, 1992 PAB meeting, distributed at the Association of Collegiate Schools of Planning Conference, Philadelphia, 1993.

15. Morris, *Planners' Salaries and Employment Trends, 1991.*

16. Ann Morrison and Kristen Crabtree, *Developing Diversity in Organizations: A Digest of Selected Literature* (Greensboro, NC: Center for Creative Leadership, 1992).

17. Roosevelt Thomas Jr., "From Affirmative Action to Affirming Diversity," *Harvard Business Review* (March-April 1990): 107-17; idem, *Beyond Race and Gender: Unleashing the Power of Your Total Work Force by Managing Diversity* (New York: American Management Association, 1991).

18. Taylor Cox Jr., "The Multicultural Organization," *Academy of Management Executive* 5 (1991): 34-47; Taylor Cox Jr. and Stacy Blake, "Managing Cultural Diversity: Implications for Organizational Competitiveness," *Academy of Management Executives* 5 (1991): 45-54.

19. James Lynch, *Multicultural Education: Principles and Practice* (London: Routledge & Kegan Paul, 1986); Joan Nordquist, *The Multicultural Education Debate in the University: A Bibliography* (Santa Cruz, CA: Reference and Research Services, 1992).

20. Gale Auletta and Terry Jones, "Reconstituting the Inner Circle," *American Behavioral Scientist* 34 (November-December 1990): 137-52.

21. James Banks, "Multicultural Education: Characteristics and Goals," in *Multicultural Education: Issues and Perspectives*, ed. James Banks and Cherry Banks (Needham Heights, MA: Allyn and Bacon, 1989). For differences by gender see Carol Gilligan, *In a Different Voice* (Cambridge, MA: Harvard University Press, 1982).

22. Banks, "Multicultural Education," 20.

23. Etta Hollins, "Debunking the Myth of a Monolithic White American Culture; or Moving Toward Cultural Inclusion," *American Behavioral Scientist* 34 (November-December 1990): 201-9; James Banks, "Integrating the Curriculum With Ethnic Content: Approaches and Guidelines," in Banks and Banks, *Multicultural Education: Issues and Perspectives.*

24. Yolanda Moses, "The Challenge of Diversity: Anthropological Perspectives on University Culture," *Education and Urban Society* 22 (August 1990): 402-12.

25. Mary Tetreault, "Integrating Content about Women and Gender into the Curriculum," in Banks and Banks, *Multicultural Education;* Banks, "Integrating the Curriculum," 198.

26. Eugenie Birch, "Woman-made America: The Case of Early Public Housing Policy," in *The American Planner: Biographies and Recollections,* ed. Don Krueckeberg (New York: Methuen, 1983); Daphne Spain, "Sustainability, Feminist Visions, and the Utopian Tradition," *Journal of Planning Literature* 9 (May 1995): 362-69.

27. St. Clair Drake and Horace Clayton, *Black Metropolis,* vol. 1: *A Study of Negro Life in a Northern City* (New York: Harcourt, Brace and World, 1970); Catherine Bauer Wurster, "Is Urban Redevelopment Possible Under Existing Legislation?" *Planning 1946* (Chicago: American Society of Planning Officials, 1946); Martin Meyerson and Edward C. Banfield, *Politics, Planning and the Public Interest* (Glencoe, IL: Free Press, 1955).

28. Eugenie Birch, "From Civic Worker to City Planner: Women and Planning, 1890-1980," in Krueckeberg, *The American Planner.*

29. Harvey Perloff and Frank Klett, "The Evolution of Planning Education," in *Learning from Turbulence,* ed. David Godschalk (Washington, DC: American Institute of Planners, 1974); Richard Klosterman, "Contemporary Planning Theory Education: Results of a Course Survey," *Journal of Planning Education and Research* 1 (1981): 1-11.

30. Sue Hendler, ed., *Planning Ethics: A Reader in Planning Theory, Practice, and Education* (New Brunswick, NJ: Center for Urban Policy Research, 1995), 125.

31. Bishwapriya Sanyal and Ibrahim Jammal, "Growing Pains of a Fruitless Endeavor? Retrospective Notes on the Buffalo Conference on Global Approaches to Planning Education," *Journal of Planning Education and Research* 11 (1992): 152-56.

32. Planning Accreditation Board, "Accreditation Criteria."

33. Banks, "Integrating the Curriculum," 192.

34. Cited in Marshall Feldman, "Perloff Revisited: Reassessing Planning Education in Postmodern Times," *Journal of Planning Education and Research* 13 (1994): 99.

35. Refer however to Marsha Ritzdorf, "Diversifying the Planning Curriculum: A Modest Bibliography," unpublished annotated bibliography. Available from Ritzdorf at Blacksburg, Virginia Polytechnic Institute and State University (1994); Lisa Servon, *The Intersection of Planning with Gender Issues,* Council of Planning Librarians Bibliography no. 303, 1993.

36. Planning Accreditation Board, "Accreditation Criteria," 1.

37. Ibid., 2.

38. Klosterman, "Contemporary Planning Theory."

39. Marsha Ritzdorf, "Feminist Thoughts on the Theory and Practice of Planning," *Planning Theory* (Newsletter) 7-8 (1992); Servon, *Intersection of Planning.*

40. American Institute of Certified Planners, *1994-95 Roster* (Washington, DC: American Institute of Certified Planners, 1994).

41. Sue Hendler, "Feminist Planning Ethics," *Journal of Planning Literature* 9 (November, 1994): 115-27; Marcia Marker Feld, "The Yonkers Case and Its Implications for the Teaching and Practice of Planning," *Journal of Planning Education and Research* 8 (1989): 169-76; Jerome Kaufman, "U.S. vs. Yonkers: A Tale of Three Planners," ibid. (1989): 189-92.

42. Jerome Kaufman, "Teaching Planning Ethics," *Journal of Planning Education and Research* 1 (1981): 33.

43. Charles Harr and Jerold Kayden, ed., *Zoning and the American Dream* (Chicago: Planners Press, 1989); Netter and Price, "Zoning and the Nouveau Poor."

44. Feld, "Yonkers Case"; see also various recent annual volumes of *Zoning and Planning Law Handbook* (New York: Clark Boardman).

45. Beth Milroy, "Taking Stock of Planning, Space, and Gender, *Journal of Planning Literature* 6 (1991): 3-15.

46. David Rusk, *Cities Without Suburbs* (Baltimore, MD: Johns Hopkins University Press, 1995).

47. Paul Davidoff and Lisa Boyd, "Peace and Justice in Planning Education," *Journal of Planning Education and Research* 3 (1983): 54.

48. Linda Dalton, "Why the Rational Paradigm Persists—The Resistance of Professional Education and Practice to Alternative Forms of Planning," *Journal of Planning Education and Research* 5 (1986): 151.

49. Hendler, "Feminist Planning Ethics"; Servon, *The Intersection*, 8-10.

50. Planning Accreditation Board, "Accreditation Criteria," 2.

51. B. Trevor Tyson and Nicholas Low, "Experiential Learning in Planning Education," *Journal of Planning Education and Research* 7 (1987): 15-28.

PART FIVE

EXCERPTS FROM SELECTED PLANNING-RELATED DOCUMENTS

A. *Village of Euclid v. Ambler Realty Corporation* (1926)

COMMENTARY

Euclid was the case that affirmed the constitutional right of communities to zone as part of the police powers that the U.S. Constitution reserves for the states, as discussed by both Silver and Ritzdorf (Chapters 2 and 3, this volume). *Euclid* allowed communities to use less obvious, but just as invidious land use decisions to segregate themselves, racially and ethnically, from the growing numbers of immigrant and in-migrant populations swelling American cities in the early years of the twentieth century. This is a selection that shows how the Court, by 1926, was already casting multifamily housing as a nonacceptable alternative for suburban communities. It is interesting to note that Justice Douglas uses very similar language in defending the need for a narrow definition in *Belle Terre* nearly forty years later (see Ritzdorf, Chapter 3, this volume). The complete case is found at 272 U.S. 394.

The matter of zoning has received much attention at the hands of commissions and experts, and the results of their investigations have been set forth in comprehensive reports. . . . With particular reference to apartment houses, it is pointed out that the development of detached house sections is greatly retarded by the coming of apartment houses, which has sometimes resulted in destroying the entire section for private house purposes; that in such sections very often the apartment house is a mere parasite, constructed in order to take advantage of the open spaces and attractive surroundings created by the residential character of the district. Moreover, the coming of one apartment house is followed by others, interfering by their height and bulk with the free circulation of air and monopolizing the rays of the sun which otherwise would fall upon the smaller homes, and bringing, as their necessary accompaniments, the disturbing noises incident to increased traffic and business, and the occupation, by means of moving and parked automobiles, of larger portions of the streets, thus detracting from their safety and depriving children of the privilege of quiet and open spaces for play, enjoyed by those in more favored localities—until, finally, the residential

character of the neighborhood and its desirability as a place of detached
residences are utterly destroyed. Under these circumstances, apartment
houses, which in a different environment would be not only entirely
unobjectionable but highly desirable, come very near to being nuisances.

B. *FHA Underwriting Manual* (1938)

COMMENTARY

The Federal Housing Administration (FHA) issued several
manuals designed to help their underwriting field staff determine
whether or not mortgage applications were eligible for insurance.
Eligibility was determined by how FHA staff rated the risks involved. The manual instructed staff how to rate several factors
related to risk, such as the nature of the property, its location, and
characteristics of the borrower. The following passages concerning
location, which clearly discouraged racial integration, guided staff
throughout the nation as they decided whether or not to approve
proposed mortgages for homes. Their biases strongly shaped suburban America since FHA helped finance much of the housing that
the real estate industry built after World War II (see Mohl, Chapter
4). Furthermore, the first passage indicates FHA's support for
restrictive covenants, which private sellers used to prevent racial
turnover (see Leavitt, Chapter 10). Selected passages come from
Federal Housing Administration, *Underwriting Manual: Underwriting and Valuation Procedure Under Title II of the National Housing Act*
(Washington, DC: U.S. Government Printing Office, 1938).

Rating of Location

- Section 902. Rating of Location has eight features, with weights assigned
  to each of them:

A. Relative Economic Stability	40
B. Protection from Adverse Influences	20
C. Freedom from Special Hazards	5

D. Adequacy of Civic, Social, and Commercial Centers ... 5
E. Adequacy of Transportation ... 10
F. Sufficiency of Utilities and Conveniences ... 5
G. Level of Taxes and Special Assessments ... 5
H. Appeal ... 10

Protection from Adverse Influences

- Section 934. *Restrictive Covenants.* Deed restrictions are apt to prove more effective than a zoning ordinance in providing protection from adverse influences. Where the same deed restrictions apply over a broad area and where the restrictions relate to types of structures, use to which improvements may be put, and occupancy, better protection is afforded. . . .
- Section 935. *Natural Physical Protection.* . . . Natural or artificially established barriers will prove effective in protecting a neighborhood and the locations within it from adverse influences. Usually the protection from adverse influences afforded by these means includes prevention of the infiltration of business and industrial uses, lower class occupancy, and inharmonious racial groups. . . .
- Section 937. *Quality of Neighborhood Development.* . . . Areas surrounding a location are investigated to determine whether incompatible racial and social groups are present, for the purpose of making a prediction regarding the probability of the location being invaded by such groups. If a neighborhood is to retain stability, it is necessary that properties shall continue to be occupied by the same social and racial classes. A change in social or racial occupancy generally contributes to instability and a decline in values.

Adequacy of Civic, Social, and Commercial Centers

- Section 951. *Quality and Accessibility of Schools.* . . . However, if the children of people living in such an area are compelled to attend school where the majority or a considerable number of the pupils represent a far lower level of society or an incompatible racial element, the neighborhood under consideration will prove far less stable and desirable than if this condition did not exist.

Appeal

• Section 973. *Social Attractiveness*. Satisfaction, contentment, and comfort result from association with persons of similar attributes. Families enjoy social relationships with other families whose education, abilities, mode of living, and racial characteristics are similar to their own. . . .

Special Considerations in Undeveloped Subdivisions:

Adequacy of Civic, Social, and Commercial Centers

• Section 982. *Adequacy of Civic, Social, and Commercial Centers.* Schools should be appropriate to the needs of the new community and they should not be attended in large numbers by inharmonious racial groups.

C. *The Urban Renewal Manual* (1961)

COMMENTARY

The following passages indicate that localities were supposed to prevent hardship and provide relocation payments, but in fact hardships were great, and payments were a mere pittance. From the beginning, wording of various redevelopment guidelines suggested that cities were supposed to be fair in their treatment of people relocated from potential redevelopment sites, and be sure that both non-Whites and Whites could be rehoused. In actuality, however, the language of agency guidelines did not protect local residents from forcible removal, or from eventual residence in unsafe substitute housing. Redevelopment caused a particular housing crisis for non-Whites, whose access to alternative lodging was limited by racial segregation. These particular passages come from a section on "Relocation" dated January 26, 1961, found in the U.S. Housing and Home Finance Agency, Urban Renewal Administration, *Urban Renewal Manual: Policies and Requirements for Local Public Agencies* (Washington, DC: U.S. Government Printing Office, various dates), 16-1.

Objectives

The objectives of the relocation program are that:

1. Families displaced by a Title I project shall have the opportunity of occupying housing that is decent, safe, and sanitary, that is within their financial means, and that is in reasonably convenient locations.
2. Displacement of site occupants shall be carried out with a minimum of hardship.

In order to assure that the displaced nonwhite families can be relocated in accordance with these objectives, information relating to housing needs and resources must be collected, analyzed, and reported by family color.

Submission Requirements

With the Survey and Planning Application, the LPA (local public agency) shall submit:

(1) Estimates of number of site occupants on Form H-6101,
(2) Brief narrative description of the housing supply in the locality, indicating white and nonwhite availability. Include the following, and identify the source of the information:
    (a) Overall number of standard and substandard housing units.
    (b) Number of private and public rental housing units.
    (c) Annual turnover rate in private and public housing units.
    (d) Vacancy ratio.
    (e) Number of private and public residential units constructed during the previous two years.
(3) Estimated costs of relocation planning, . . .
(4) Amount, basis, and source of estimates for Section 106(f) Relocation Payments.

D. *The Racial Aspects of Urban Planning:* An Urban League Research Report (1968)

COMMENTARY

By the mid-1960s the African American community was no longer silent about the racial injustices of urban planning and of various federal policies, particularly the urban renewal program. The following passages represent one of the most complete statements available from a civil rights organization. Excerpts come from Baron's introduction to Harold M. Baron, ed., *The Racial Aspects of Urban Planning: Critique on the Comprehensive Plan of the City of Chicago, with Commentaries* (Chicago: Chicago Urban League, 1968), 8-12. Copyright 1968, Chicago Urban League. Used with permission.

As the metropolitan centers grow in size and become more technologically complex, the techniques of planning for them have become more methodologically sophisticated. The layman stands in awe of the means employed to make planning more comprehensive and comprehensible. Yet, the application of computer technology is still in its infancy. So far it has been more extensively employed in transportation than land usage studies, but large-scale data banks for land use information will soon be operational. Nevertheless, all this professional sophistication seems far removed from the number one problem of the city, racism. The aware black man in the ghetto tends to view urban planning more as an enemy than as an aid. There should be no wonder in this, for public housing monstrosities and urban renewal, with its callous treatment in the displacement of Negro householders, are the two most obvious and familiar evidences of planning to him.

The major traditions in the profession of planning emphasize creating physical order in spatial arrangements and serving as technical advisers to those placed in power by the given political and quasi-political processes. Planners, therefore, have been the most successful in those cases where there has been little effective political opposition and where they have been dealing with fairly specific planning problems. Operating under the concept of the general good as defined by the political consensus of the moment, planners have rarely stopped to ask, "*who* is sacrificing what for *whom*?" It is assumed that the subject and the object of the action are part of an indistinguishable general public without basically divided interests; or, if someone's interest must be sacrificed, it is an interest in property that will be properly protected in the courts through procedures in equity.

When a potential opposition is powerless and property-less (are not these terms almost synonymous in our society?), as is the case of Negroes in the city, the question of who gives up what becomes of utmost importance. The specific problems that are eventually dealt with, and for which plans get implemented, tend to be those which represent the interests of prestigious persons or prestigious institutions. Urban renewal offers an excellent illustration of the underlying conflict in present-day planning programs that are being carried out under the banner of the general public good. Removing slums and upgrading the environment of prestigious institutions, such as universities and hospitals or central business districts, has been the major result of urban renewal programs. . . .

Urban racism is based upon a web of institutional arrangements that permeate the basic social structure of the city. The planner, therefore, is confronted with the fact that racial oppression is embodied not only in the physical presence of the ghetto, but also in many institutional structures that extend throughout the metropolitan region. Through the studies at the Chicago Urban League we have come to the conclusion that racism is institutionalized in the urban environment through the existence of well defined and subordinated Negro subsectors of such major institutional areas as the labor market, the educational system, the housing market and the political structure. The outcomes from any one of these areas supports the discrimination that exists in the others so that there is a mutually supportive web of urban racism in which the whole is greater than the sum of the parts.

Social scientists have paid little attention to the study of these racist institutional structures; accordingly, they have provided the planner with very little to work with on this line. When, for example, little is known about the nature of the racially dual labor market except that discrimination exists, it is very difficult for the planner, even if he has the will, to figure out the racial implication of different physical arrangements or residences and places of employment.

While social scientists are to be rightly faulted for not providing many interpretative understandings about racist institutions in the city, there is great question about how well prepared planners would be to use this information if they had it. First, they would have to find the means of overcoming the limitations in the outlook of the official bodies that they serve so that they could realistically propose humane rather than oppressive uses of this information. Second, planners have been most successful in dealing with social information where there are statistical regularities involving large numbers, such as: market behavior, transportation patterns, or demographic information on births, deaths, and mobility. They have been less successful in dealing with the broad range of institutional elements where uniqueness has to be taken into account. In these cases, they tend to give full recognition only to those institutions that are powerful in defense of their own interests.

The planner's lot is not a happy one. Politically, he is hemmed in; for at this stage of our history, even if the planner could devise a scheme which would provide the kind of physical environment conducive to the erosion of racism, Negroes and their allies do not have sufficient power to get the plans adopted. The Chicago Urban League, for example, published *A Plan for a System of Educational Parks*, which would provide a physical environment very conducive to breaking down racism in the educational system of Chicago; yet, little serious consideration has been given its desirability or practicality.

The affluent suburbs are given almost a free hand in controlling their "turfs"; they have self-determination in its most unvarnished form. Large prestigious institutions can develop their own planning capabilities and muster the political strength to get a good deal of what they want. But, if you are poor and/or black in the large central city, you take what is given to you. Whether the giving is malicious or paternalistic, you are not involved in the determination of the plans. This situation forces the planner to become a partner of the powerful and a manipulator of the masses.

Two hopeful developments appear as though they might partially resolve this latter dilemma. In many local communities and in the planning profession itself, there has developed a program of advocacy planning. Local groups, especially in existing neighborhoods expecting conservation or redevelopment, engage their own planners to develop schemes of land-use and building design which will reflect the needs and interests of the low-prestige persons living there. Advocacy planning operates as a counter-weight to the work of the large city's own planning staff which, out of current political necessity, is much more sensitive to the interests of the more prestigious persons and institutions. The presentation of workable alternatives to the local communities gives them both hope and the instrumentality by which they can become active in their own behalf, rather than passively accepting as fate the decisions made in the interest of those already powerful. This point leads to the second hopeful development: the increasing political power of the black people, who, because they are locked in the central cities, are rapidly growing as a proportion of the population. This political shift could well provide planners a different kind of constituted authority to serve. If these new authorities do their job correctly, they will demand of the planners a new kind of vision.

The Chicago Urban League concluded a few years back that one of its major technical functions in helping to eradicate racism would be to make a start at unravelling the racial mysteries of urban planning.

E. *Village of Belle Terre v. Boraas* (1973)

COMMENTARY

As discussed by Ritzdorf in Chapters 3 and 5, the decision in this case allows municipal zoning officials the right to regulate the intimate composition of households in their community. In his dissent, portions of which are reprinted below, Justice Marshall vehemently disagrees with his colleagues on the constitutionality of such actions. He challenges their logic about the parameters of the constitutionally protected right to privacy. He clearly points out the fallibility of a planning logic that claims it is protecting the family character of its neighborhoods by allowing a family of twelve to live in a three-room house while denying three elderly people the right to share a mansion. The complete case and dissent are found at 416 U.S. 1.

Mr. Justice Marshall, dissenting:

Zoning officials properly concern themselves with the uses of land—with, for example, the number and kind of dwellings to be constructed in a certain neighborhood or the number of persons who can reside in those dwellings. But zoning authorities cannot validly consider who those persons are, what they believe, or how they choose to live, whether they are Negro or white, Catholic or Jew, Republican or Democrat, married or unmarried.

My disagreement with the Court today is based upon my view that the ordinance in this case unnecessarily burdens appellees' First Amendment freedom of association and their constitutionally guaranteed right to privacy. . . . Constitutional protection is extended, not only to modes of association that are political in the usual sense, but also to those that pertain to the social and economic benefit of the members. . . . The selection of one's living companions involves similar choices as to the emotional, social or economic benefits to be derived from alternative living arrangements.

The choice of household companions—of whether a person's "intellectual and emotional" needs are best met by living with family, friends, professional associates or others—involves deeply personal considerations as to the kind and quality of intimate relationships within the home. That decision surely falls within the ambit of the right to privacy protected by the Constitution. . . .

This ordinance discriminates on the basis of just such a personal lifestyle choice as to household companions. It permits any number of persons related by blood or marriage, be it two or twenty, to live in a single household, but it limits to two the number of unrelated persons bound by profession, love, friendship, religious or political affiliation, or mere economics who can occupy a single home. *Belle Terre* imposes upon those who deviate from the community norm in their choice of living companions significantly greater restrictions than are applied to residential groups who are related by blood or marriage, and compose the established order within the community. The village has, in effect, acted to fence out those individuals whose choice of lifestyle differs from that of its current residents. . . .

A variety of justifications have been proffered in support of the village's ordinance. It is claimed that the ordinance controls population density, prevents noise, traffic and parking problems, and preserves the rent structure of the community and its attractiveness to families. As I noted earlier, these are all legitimate and substantial interests of government. But I think it clear that the means chosen to accomplish these purposes are both overinclusive and underinclusive, and that the asserted goals could be as effectively achieved by means of an ordinance that did not discriminate on the basis of constitutionally protected choices of lifestyle. The ordinance imposes no restriction whatsoever on the number of persons who may live in a house, as long as they are related by marital or sanguinary bonds—presumably no matter how distant their relationship. Nor does the ordinance restrict the number of income earners who may contribute to rent in such a household, or the number of automobiles that may be maintained by its occupants. In that sense the ordinance is underinclusive. On the other hand, the statute restricts the number of unrelated persons who may live in a home to no more than two. It would therefore prevent three unrelated people from occupying a dwelling even if among them they had but one income and no vehicles. While an extended family of a dozen or more might live in a small bungalow, three elderly and retired persons could not occupy the large manor house next door. Thus the statute is also grossly overinclusive to accomplish its intended purposes. . . .

Appellants also refer to the necessity of maintaining the family character of the village. There is not a shred of evidence in the record indicating that if Belle Terre permitted a limited number of unrelated persons to live together, the residential, familial character of the community would be fundamentally affected. By limiting unrelated households to two persons while placing no limitation on households of related individuals, the village has embarked upon its commendable course in a constitutionally faulty vessel. . . .

I respectfully dissent. (416 U.S. 14-20)

F. *Warth v. Seldin* (1974)

COMMENTARY

The right to bring a case before the court is called *standing*. When the Supreme Court used the *Warth* case to deny standing to sue to a set of plaintiffs who wanted to bring low and moderate income housing to Penfield, they effectively closed the door on suits that would aid the opening of suburban communities to low income and minority residents (see Ritzdorf, Chapter 3, this volume). The plaintiffs represented three constituencies: potential residents, builders, and a local nonprofit housing corporation. These selections come from the dissents of Justices Douglas and Brennan. Justice Brennan, especially, clearly recognizes and exposes the issues of race and class that framed the Court's response to the case before them. The complete case can be found at 422 U.S. 490.

Mr. Justice Douglas, dissenting:

With all respect I think that the Court reads the complaint and the record with antagonistic eyes. There are in the background of this case continuing strong tides of opinion touching on very sensitive matters, some of which involve race, some class distinctions based on wealth. . . .

Standing has become a barrier to access to the federal courts, just as "the political question" was in earlier decades. The mounting caseload of federal courts is well known. But cases such as this one reflect festering sores in our society; and the American dream teaches that if one reaches high enough and persists there is a forum where justice is dispensed. I would lower the technical barriers and let the courts serve that ancient need. . . .

As Mr. Justice Brennan makes clear in his dissent, the alleged purpose of the ordinance under attack was to preclude low- and moderate-income people and non-whites from living in Penfield. The zoning power is claimed to have been used here to foist an un-American community model on the people of this area. I would let the case go to trial and have all the facts brought out. Indeed, it would be better practice to decide the question of standing only when the merits have been developed. I would reverse the Court of Appeals. (422 U.S. 518-19)

Mr. Justice Brennan, with whom Mr. Justice White and Mr. Justice Marshall join, dissenting:

In this case, a wide range of plaintiffs, alleging various kinds of injuries, claimed to have been affected by the Penfield zoning ordinance on its face and as applied and by other practices of the defendant officials of Penfield. Alleging that as a result of these laws and practices low- and moderate-income and minority people have been excluded from Penfield, and that this exclusion is unconstitutional, plaintiffs sought injunctive, declaratory, and monetary relief. The Court today, in an opinion that purports to be a "standing" opinion but that actually, I believe, has overtones of outmoded notions of pleading and of justifiability, refuses to find that any of the variously situated plaintiffs can clear numerous hurdles, some constructed here for the first time, necessary to establish "standing." While the Court gives lip service to the principle, oft repeated in recent years, that "standing in no way depends on the merits of the plaintiffs contention that particular conduct is illegal," in fact the opinion which tosses out of court almost every conceivable kind of plaintiff who could be injured by the activity claimed to be unconstitutional, can be explained only by an indefensible hostility to the claim on the merits. I can appreciate the Court's reluctance to adjudicate the complex and difficult legal questions involved in determining the constitutionality of practices which assertedly limit residence in a particular municipality to those who are white and relatively well off, and I also understand that the merits of this case could involve grave sociological and political ramifications. But courts cannot refuse to hear a case on the merits merely because they would prefer not to, and it is quite clear, when the record is viewed with dispassion, that at least three of the groups of plaintiffs have made allegations, and supported them with affidavits and documentary evidence, sufficient to survive a motion to dismiss for lack of standing. . . .

Accepting, as we must, the various allegations and affidavits as true, the following picture emerges: The Penfield zoning ordinance, by virtue of regulations concerning "lot area, set backs, . . . population density,

density of use, units per acre, floor area, sewer requirements, traffic flow, ingress and egress and street location," makes "practically and economically impossible the construction of sufficient numbers of low and moderate income" housing. The *purpose* of this ordinance was to preclude low- and moderate-income people and nonwhites from living in Penfield.

As a result of these practices, various of the plaintiffs were affected in different ways. For example, plaintiffs Ortiz, Reyes, Sinkler, and Broadnax, persons of low or moderate income and members of minority groups, alleged that "as a result" of respondents' exclusionary scheme, they could not live in Penfield, although they desired and attempted to do so, and consequently incurred greater commuting costs, lived in substandard housing, and had fewer services for their families and poorer schools for their children than if they had lived in Penfield. Members of the Rochester Home Builders Association were prevented from constructing homes for low- and moderate-income people in Penfield, harming them economically. And Penfield Better Homes, a member of the Housing Council, was frustrated in its attempt to build moderate-income housing.

Thus, the portrait which emerges from the allegations and affidavits is one of total, purposeful, intransigent exclusion of certain classes of people from the town, pursuant to a conscious scheme never deviated from. Because of this scheme, those interested in building homes for the excluded groups were faced with insurmountable difficulties, and those of the excluded groups seeking homes in the locality quickly learned that their attempts were futile. . . .

Understandably, today's decision will be read as revealing hostility to breaking down even unconstitutional zoning barriers that frustrate the deep human yearning of low-income and minority groups for decent housing they can afford in decent surroundings. (422 U.S. 519-29)

G. *New Federalism and Community Development: A Joint Center for Political Studies Report (1976)*

COMMENTARY

The Joint Center for Political and Economic Studies is a non-profit organization that undertakes research that often concerns African Americans and political issues. In 1976, two years after passage of the act that established Community Development Block Grants (CDBGs), they published a commentary that proved remarkably prescient concerning the probable effects of CDBGs on minority groups. The following excerpts come from pages 35-36 of that report, authored by Milton Morris, *New Federalism and Community Development: A Preliminary Evaluation of The Housing and Community Development Act of 1974* (Washington, DC: Joint Center for Political Studies, 1976). The report begins by explaining New Federalism and the nature of the CDBG program. Chapter 3, here cited, focuses on strategies that minority groups must undertake in the policy climate of the 1970s, which was very different from the era of the War on Poverty and Model Cities (see Thomas, Chapter 9, this volume). Copyright Joint Center for Political Studies, 1976. Used with permission.

[New Federalism] reflects a strong desire to alter or eliminate those programs and implementation strategies widely regarded as "liberal" innovations of the 1960s. The community development [CDBG] program must be seen, in part at least, as a major part of this retrenchment effort.

Two types of changes inaugurated by the new program reflect this effort at retrenchment. The first involves the concentration of responsibility for all community development activities on general-purpose local governments—city halls—in contrast to the practice of utilizing special purpose bodies for some task and the extensive involvement of the federal bureaucrats in determining the choice of activities and in supervising their implementation. The Act limits HUD primarily to reviewing applications to ensure compliance with broad legislative provisions, apportioning most funds by standard formula, and reviewing annual performance reports submitted by local governments.

apportioning most funds by standard formula, and reviewing annual performance reports submitted by local governments.

The second change derives from the utilization of a fixed formula for distributing funds which broadens substantially the types of communities which will benefit from community development funds. The urban renewal program focused on dilapidated inner cities, and Model Cities was even more precisely targeted—at particularly dilapidated neighborhoods in selected cities. Although the Act tries to retain the focus on low- and moderate-income groups, in actuality the block grant program extends these funds into suburbia and to scores of small cities without treating redevelopment needs.

These changes have substantial implications for disadvantaged minority groups. They must now seek to protect their interests in community development activities by influencing local governments instead of concentrating almost entirely on federal officials. The potential difficulties in this kind of effort appear staggering, especially when the historical record of state and local governments with respect to minority groups like blacks is considered. Past experience provides overwhelming evidence that in the absence of strong federal involvement blacks in particular have been the victims of widespread indifference or active hostility by many state and local governments. It was partly because of this indifference and unresponsiveness that federal policy-makers in the 1960s sought to devise programs which bypassed state and local governments and went directly to needy and inadequately served communities populated in most cases by minorities.

Within recent years there has been more direct minority participation in local governments and consequently these governments have done more to meet the needs of minority groups. There is, however, little evidence of the widespread and far-reaching changes that would make these groups optimistic about how they will fare with a drastically reduced federal presence in local community development activities.

For blacks in particular, the most significant factor operating in their behalf is the greater access they now have to the political arena at all levels of government. Their response to this changing pattern of federalism must be greatly intensified political activity focusing on all areas of local government. The transfer of large amounts of federal funds and new responsibilities to local governments through this program and the revenue sharing program means that blacks can no longer ignore city hall in favor of Washington. Furthermore, it means that the extent to which blacks benefit from programs like community development will be directly related to their skill and persistence as participants at the local level and to a much lesser extent on individuals and organizations operating at the national level.

H. U.S. Fair Housing Act (1988)

COMMENTARY

These selections are from the federal Fair Housing Act as amended in 1988. The first section explains the often confusing exceptions from its application (which applies to much available rental housing) since many rental units are in buildings where the owner qualifies to be exempted. As a result, the exceptions provide another, subtle way in which the stock of housing available to low-income and minority renters and owners is limited. The second section describes who is considered to be in the real estate business. The third section describes the scope of the protection of the Act after the 1988 additions regarding familial status (this refers to households containing children) and the rights of the handicapped. The complete act can be found at United States Code, 1989, Title 42, chapter 45, sections 3601-3631.

---

§3603 *(b) Exemptions*

Nothing in section 3604 of this title (other than subsection (c)) shall apply to—

1. any single-family house sold or rented by an owner: *Provided,* That such private individual owner does not own more than three such single family houses at any one time. *Provided further,* That in the case of the sale of any such single-family house by a private individual owner not residing in such house at the time of such sale or who was not the most recent resident of such house prior to such sale, the exemption granted

by this subsection shall apply only with respect to one such sale within any twenty-four month period; *Provided further,* That such bona fide private individual owner does not own any interest in, nor is there owned or reserved on his behalf, under any express or voluntary agreement, title to or any right to all or a portion of the proceeds from the sale or rental of more than three such single-family houses at any one time; *Provided further,* That after December 31, 1969, the sale or rental of any such single-family house shall be excepted from the application of this subchapter only if such house is sold or rented (A) without the use in any manner of the sales or rental facilities or the sales or rental services of any real estate broker, agent, or salesman, or of such facilities or services of any person in the business of selling or renting dwellings, or of any employee or agent of any such broker, agent, salesman, or person and (B) without the publication, posting or mailing, after notice, of any advertisement or written notice in violation of section 3604(c) of this title; but nothing in this proviso shall prohibit the use of attorneys, escrow agents, abstractors, title companies, and other such professional assistance as necessary to perfect or transfer the title, or

2. rooms or units in dwellings containing living quarters occupied or intended to be occupied by no more than four families living independently of each other, if the owner actually maintains and occupies one of such living quarters as his residence.

(c) Business of Selling or Renting Dwellings Defined

For the purposes of subsection (b) of this section, a person shall be deemed to be in the business of selling or renting dwellings if—

1. he has, within the preceding twelve months, participated as principal in three or more transactions involving the sale or rental of any dwelling or any interest therein, or

2. he has, within the preceding twelve months, participated as agent, other than in the sale of his own personal residence in providing sales or rental facilities or sales or rental services in two or more transactions involving the sale or rental of any dwelling or any interest therein, or

3. he is the owner of any dwelling designed or intended for occupancy by, or occupied by, five or more families. (Pub. L. 90-284, Title VIII, 803, Apr. 11, 1968, 82 Stat. 82.)

§3604. Discrimination in the Sale or Rental of Housing and Other Prohibited Practices

As made applicable by section 3603 of this title and except as exempted by sections 3603(b) and 3607 of this title, it shall be unlawful—

(a) To refuse to sell or rent after the making of a bona fide offer, or to refuse to negotiate for the sale or rental of, or otherwise make unavailable or deny, a dwelling to any person because of race, color, religion, sex, familial status, or national origin.

(b) To discriminate against any person in the terms, conditions, or privileges of sale or rental of a dwelling, or in the provision of services or facilities in connection therewith, because of race, color, religion, sex, familial status, or national origin.

(c) To make, print, or publish, or cause to be made, printed, or published any notice, statement, or advertisement, with respect to the sale or rental of a dwelling that indicates any preference, limitation, or discrimination based on race, color, religion, sex, handicap, familial status, or national origin, or an intention to make any such preference, limitation, or discrimination.

(d) To represent to any person because of race, color, religion, sex, handicap, familial status, or national origin that any dwelling is not available for inspection, sale, or rental when such dwelling is in fact so available.

(e) For profit, to induce or attempt to induce any person to sell or rent any dwelling by representations regarding the entry or prospective entry into the neighborhood of a person or persons of a particular race, color, religion, sex, handicap, familial status, or national origin.

(f) (1) To discriminate in the sale or rental, or to otherwise make unavailable or deny, a dwelling to any buyer or renter because of a handicap of—(A) that buyer or renter; (B) a person residing in or intending to reside in that dwelling after it is so sold, rented, or made available; or (C) any person associated with that buyer or renter.

(2) To discriminate against any person in the terms, conditions, or privileges of sale or rental of a dwelling, or in the provision of services or facilities in connection with such dwelling, because of a handicap of—(A) that person; or (B) a person residing in or intending to reside in

that dwelling after it is so sold, rented, or made available; or (C) any person associated with that person.

(3) For purposes of this subsection, discrimination includes—(A) a refusal to permit, at the expense of the handicapped person, reasonable modifications of existing premises occupied or to be occupied by such person if such modifications may be necessary to afford such person full enjoyment of the premises except that, in the case of a rental, the landlord may where it is reasonable to do so condition permission for a modification on the renter agreeing to restore the interior of the premises to the condition that existed before the modification, reasonable wear and tear excepted; (B) a refusal to make reasonable accommodations in rules, policies, practices, or services, when such accommodations may be necessary to afford such person equal opportunity to use and enjoy a dwelling; or (C) in connection with the design and construction of covered multifamily dwellings for first occupancy after the date that is 30 months after September 13, 1988, a failure to design and construct those dwellings in such a manner that the public use and common use portions of such dwellings are readily accessible to and usable by handicapped persons.

I. *American Institute of Certified Planners Code of Ethics* (1989)

COMMENTARY

The following are excerpts from the American Institute of Certified Planners (AICP) Code of Ethics and Professional Conduct, as adopted April 29, 1989. AICP is the professional organization of U.S. urban planners who have passed a written professional competency examination and have practiced in the planning field for a minimum number of years. This code shows how the profession has begun to respond to the need for social equity regarding race and poverty, as called for in several chapters, particularly in Krumholz, Chapter 7, this volume. Thomas (Chapter 15, this volume) suggests that classes in planning theory and ethics should carefully study these and similar provisions of the professional ethics code, available from the preliminary pages of any current *AICP Member-ship Roster* (Washington, DC: AICP).

A.4. A planner must strive to give citizens the opportunity to have a meaningful impact on the development of plans and programs. Partici-pation should be broad enough to include people who lack formal organization or influence.

A.5. A planner must strive to expand choice and opportunity for all persons, recognizing a special responsibility to plan for the needs of disadvantaged groups and persons, and must urge the alteration of policies, institutions and decisions which oppose such needs. . . .

C.7. A planner must strive to increase the opportunities for women and members of recognized minorities to become professional planners. . . .

D.2. A planner must respect the rights of others and, in particular, must not improperly discriminate against persons. . . .

D.6. A planner must strive to contribute time and effort to groups lacking in adequate planning resources and to voluntary professional activities.

J. President Clinton's "Environmental Justice" Executive Order (1994)

COMMENTARY

As discussed by Collin and Collin (Chapter 13, this volume), this executive order represents a concrete move by President Clinton toward a recognition of the need to address the environmental needs of low income and minority populations. As with any execu-tive order, however, its impact can be very limited, as the EPA is only bound to enforce it while he is president. However, it is an excellent example of the power of steady grassroots activism to move forward a political agenda. It is because of the work of urban and minority environmental groups that this order was conceived and implemented. This is an abridged version; for the complete version, see 59 FR 7629, 1994 WS 43891 (Pres.).

EXECUTIVE ORDER 12898: Federal Actions To Address Environmental Justice in Minority Populations and Low-Income Populations, February 11, 1994

By the authority vested in me as President by the Constitution and the laws of the United States of America, it is hereby ordered as follows:

Section 1-1. Implementation.

1-101. *Agency Responsibilities.* To the greatest extent practicable and permitted by law, and consistent with the principles set forth in the report on the National Performance Review, each Federal agency shall make achieving environmental justice part of its mission by identifying and addressing, as appropriate, disproportionately high and adverse human health or environmental effects of its programs, policies, and activities on minority populations and low-income populations in the United States and its territories and possessions, the District of Columbia, the Commonwealth of Puerto Rico, and the Commonwealth of the Mariana Islands.

1-102. *Creation of an Interagency Working Group on Environmental Justice.* (a) Within 3 months of the date of this order, the Administrator of the Environmental Protection Agency ("Administrator") or the Administrator's designee shall convene an interagency Federal Working Group on Environmental Justice ("Working Group"). . . . The Working Group shall report to the President through the Deputy Assistant to the President for Environmental Policy and the Assistant to the President for Domestic Policy. (b) The Working Group shall: (1) provide guidance to Federal agencies on criteria for identifying disproportionately high and adverse human health or environmental effects on minority populations and low income populations; (2) coordinate with, provide guidance to, and serve as a clearinghouse for each Federal agency as it develops an environmental justice strategy as required by section 1-103 of this order, in order to ensure that the administration, interpretation and enforcement of programs, activities and policies are undertaken in a consistent manner; (3) assist in coordinating research by, and stimulating cooperation among, the Environmental Protection Agency, the Department of Health and Human Services, the Department of Housing and Urban Development, and other agencies conducting research or other activities in accordance with section 3-3 of this order; (4) assist in coordinating data collection, required by this order; (5) examine existing data and studies on environmental justice; (6) hold public meetings as required in section 5-502(d) of this order; and (7) develop interagency model projects on environmental justice that evidence cooperation among Federal agencies.

1-103. *Development of Agency Strategies.* (a) Except as provided in section 6-605 of this order, each Federal agency shall develop an agency-wide environmental justice strategy, as set forth in subsections (b)-(e) of this section that identifies and addresses disproportionately high and adverse human health or environmental effects of its programs, policies, and activities on minority populations and low-income populations. The environmental justice strategy shall list programs, policies, planning and public participation processes, enforcement, and/or rulemakings related to human health or the environment that should be revised to, at a minimum: (1) promote enforcement of all health and environmental statutes in areas with minority populations and low-income populations; (2) ensure greater public participation; (3) improve research and data collection relating to the health of and environment of minority populations and low-income populations; and (4) identify differential patterns of consumption of natural resources among minority populations and low-income populations. In addition, the environmental justice strategy shall include, where appropriate, a timetable for undertaking identified revisions and consideration of economic and social implications of the revisions. . . .

Sec. 2-2. Federal Agency Responsibilities for Federal Programs.

Each Federal agency shall conduct its programs, policies, and activities that substantially affect human health or the environment, in a manner that ensures that such programs, policies, and activities do not have the effect of excluding persons (including populations) from participation in, denying persons (including populations) the benefits of, or subjecting persons (including populations) to discrimination under, such programs, policies, and activities, because of their race, color, or national origin.

Sec. 3-3. Research, Data Collection, and Analysis.

3-301. *Human Health and Environmental Research and Analysis.* (a) Environmental human health research, whenever practicable and appropriate, shall include diverse segments of the population in epidemiological and clinical studies, including segments at high risk from environmental hazards, such as minority populations, low-income populations and workers who may be exposed to substantial environmental hazards. (b) Environmental human health analyses, whenever practicable and appropriate, shall identify multiple and cumulative exposures. (c) Federal agencies shall provide minority populations and low-income populations the opportunity to comment on the development and design of research strategies undertaken pursuant to this order.

3-302. *Human Health and Environmental Data Collection and Analysis.* To the extent permitted by existing law, including the Privacy Act, as amended (5 U.S.C. section 552a): (a) each Federal agency, whenever practicable and appropriate, shall collect, maintain, and analyze information assessing and comparing environmental and human health risks borne by populations identified by race, national origin, or income. To the extent practical and appropriate, Federal agencies shall use this information to determine whether their programs, policies, and activities have disproportionately high and adverse human health or environmental effects on minority populations and low-income populations. . . .

WILLIAM CLINTON, THE WHITE HOUSE, February 11, 1994. Exec. Order No. 12898, 59 FR 7629, 1994 WL 43891 (Pres.).

K. *Community Reinvestment Act: Challenges Remain* (1995)

COMMENTARY

The following excerpts come from an important study issued in 1995 by the federal "watchdog" agency, the General Accounting Office (GAO), concerning the Community Reinvestment Act (CRA). Some observers believe that CRA is as important an "urban policy" as the more visible Empowerment Zone/Enterprise Community legislation (described briefly in Thomas's Chapter 9), because of its role in spurring inner-city investment. In fact, CRA helps localities that have received EZ/EC obtain private funds, because it encourages lending institutions to invest in designated distressed areas. Changes over two decades in CRA and related mortgage disclosure legislation have begun to attack the pernicious practice by which lending institutions "redlined," or refused to invest, in inner-city areas, and by which they discriminated against racial minorities. While legislative revisions in the 1990s have been particularly helpful, however, this GAO report suggests that problems remain even with new 1995 regulations.

The first two sections of the following material come from pages 17-20 of the GAO report, and summarize CRA initiatives and the implications for fair housing. The remainder sections come from the Executive Summary, pages 2-5. The reader may order a full copy of the report—"Community Reinvestment Act: Challenges Remain to Successfully Implement CRA" (Washington, DC: U.S. General Accounting Office, November 1995)—by contacting the GAO and citing the report's title or access number (GAO/GGD-96-23).

Introduction

The Community Reinvestment Act (CRA) was passed as Title VIII of the Housing and Community Development Act of 1977 (12 U.S.C. 2901 et seq.). CRA requires each federal banking regulator to use its authority, when examining institutions, to encourage such institutions to help meet the credit needs of the local communities in which they are chartered, consistent with the institution's safe and sound operation. In connection with these examinations, the regulators are required to assess an institution's record of lending in its community and take it into account when evaluating any type of application by an institution for a deposit facility.

CRA was amended by FIRREA [Financial Institutions Reform, Recovery, and Enforcement Act] to require that the regulator's examination rating and a written evaluation of each assessment factor be made publicly available. FIRREA also established a four-part qualitative rating scale so that the publicly available CRA ratings would not be confused with the five-part numerical ratings given to institutions by the regulators on the basis of the safety and soundness of their operations. These safety and soundness ratings are confidential. In 1991, FDICIA [Federal Deposit Insurance Corporation Improvement Act] further amended CRA to require public discussion of data underlying the regulators' assessment of an institution's CRA performance in the public CRA evaluation. The Housing and Community Development Act of 1992 amended CRA to require that the regulators consider activities and investments involving minority- and women-owned financial institutions and low-income credit unions in assessing the CRA performance of institutions cooperating in these efforts. The Riegle-Neal Interstate Banking and Branching Efficiency Act of 1994 amended CRA to require that institutions with interstate branching structures receive a separate rating and written evaluation for each state in which they have branches and a separate written evaluation of their performance within a multi-state metropolitan area where they have branches in two or more states within the area.

The principle contained in CRA, that institutions must serve the "convenience and needs" of the communities in which they are chartered to do business consistent with safe and sound operations, is one that federal law governing deposit insurance, bank charters, and bank mergers had embodied before CRA was enacted. The Banking Act of 1935 declared that banks should serve the convenience and needs of their communities. The Bank Holding Company Act, initially passed in 1956, requires FRB [Federal Reserve Board], in acting on acquisitions by banks and bank holding companies, to evaluate how well a bank meets the convenience and needs of its communities within the limits of safety and soundness. Under CRA, the concept of "convenience and needs" was refined to explicitly include extensions of credit.

The Fair Housing Laws
Are Related to CRA

CRA and the fair lending laws, while separate, have related objectives. The primary purpose of CRA was to prohibit redlining—arbitrarily failing to provide credit to low- and moderate-income neighborhoods. FHA [Fair Housing Act] and ECOA [Equal Credit Opportunity Act] prohibit lending discrimination based on certain characteristics of potential and actual borrowers. The FHA, passed by Congress in 1968 as Title

VIII of the Civil Rights Act of 1968, among other things prohibits discrimination in residential real estate-related transactions on the basis of an applicant's race, color, religion, gender, handicap, familial status, or national origin. Such prohibited activities include denying or fixing the terms and conditions of a loan based on discriminatory criteria. The ECOA, passed in 1974, prohibits discrimination with respect to any aspect of a credit transaction based on race, color, religion, national origin, gender, marital status, age, receipt of public assistance, or the exercise, in good faith, of rights granted by the Consumer Credit Protection Act.

[The Home Mortgage Disclosure Act (HMDA)] was enacted by Congress in 1975 to provide regulators and the public with information so that both could determine whether depository institutions were serving the credit needs of their communities but was expanded over time to detect evidence of possible discrimination based on the individual characteristics of applicants. HMDA established a reporting obligation for depository institutions. Initially, HMDA required depository institutions with total assets of more than $10 million to compile data on the number and total dollar amount of mortgage loans originated or for which the institution received completed applications or purchased during each fiscal year by geographic area and make that data available for public inspection. In 1989, HMDA was amended to require collection and reporting of data on race, gender, and income characteristics of mortgage applicants to provide data to assist in identifying discriminatory lending practices and enforcing fair lending statutes. Amendments to HMDA in 1988 and 1991 expanded the reporting requirements to most mortgage banking subsidiaries of bank and thrift holding companies and independent mortgage companies not affiliated with depository institutions. In 1992, HMDA was amended to require affected financial institutions to make available to the public, upon request, their loan application registers, which maintain data for loans covered by HMDA.

Both HMDA and CRA were originally enacted to remedy a perceived lack of lending by institutions to the communities in which they were chartered to do business by the regulators. HMDA was amended in 1989 to include the collection of data on race, sex, and income of applicants for credit to provide indications of possible lending discrimination. In addition, 2 of the 12 assessment factors, factors D and F, in the current CRA regulation address the issue of discrimination to be considered in determining an institution's CRA rating. Where available, HMDA data are to be used by examiners when assessing compliance with CRA, FHA, and ECOA.

Background of 1995 GAO Study

The former Chairmen, House Committee on Banking, Finance and Urban Affairs and its Subcommittee on Consumer Credit and Insurance, asked GAO to address four questions: (1) What were the major problems in implementing CRA, as identified by the affected parties—bankers, regulators, and community groups? (2) To what extent do the regulatory reforms address these problems? (3) What challenges do the regulators face in ensuring the success of the reforms and what, if any, actions would help the regulators in facing these challenges? and (4) What initiatives have been taken or proposed to help bankers overcome community lending barriers and enhance lending opportunities, particularly in low- and moderate-income areas?

The debate preceding enactment of CRA was similar to the current debate. Community groups urged its passage to curb what they believed to be a lack of adequate lending in low- and moderate-income areas. Bankers generally opposed CRA as an unnecessary measure that could adversely affect business decisions by mandating credit allocation and cause safety and soundness problems by forcing institutions to make excessively risky loans. More recently, changing market conditions along with increased public disclosure have raised bankers' concerns about the issues of competition and regulatory burden. More specifically, bankers have become concerned about the competitive advantages for nonbank financial institutions, such as mortgage companies, that compete with banks but are not subject to CRA requirements. Bankers also objected that the cost and paperwork burdens imposed by CRA are not offset by positive incentives, such as protection against protests of expansion plans, to encourage CRA compliance. However, community groups have raised concerns about limited CRA enforcement and insufficient disclosure of information on institutions' community lending performance.

As concerns about CRA increased from all affected parties, both the administration and Congress looked for ways to make CRA more effective and less burdensome. The stated goals of the regulators' reform initiative, announced by the President in July 1993, were to (1) base CRA examinations more on results than paperwork, (2) clarify performance standards, (3) make examinations more consistent, (4) improve enforcement to provide more effective sanctions, and (5) reduce the cost and burden of compliance. Subsequently, the regulators issued two notices of proposed rule-making and, after receiving extensive public comments, promulgated the revised CRA regulations in May 1995. Several legislative proposals have also sought to reduce the burden associated with CRA compliance.

Results in Brief:
GAO Monitoring Study

Through interviews with bankers, community groups, and regulatory officials, GAO identified four major problems with the regulators' compliance examinations and enforcement of CRA that all the affected parties agreed were problems: (1) too little reliance on lending results and too much reliance on documentation of efforts and processes, leading to an excessive paperwork burden; (2) inconsistent CRA examinations by regulators resulting in uncertainty about how CRA performance is to be rated; (3) examinations based on insufficient information that may not reflect a complete and accurate measure of institutions' performance; and (4) dissatisfaction with regulatory enforcement of the act, which largely relies on protests of expansion plans to ensure institutions are responsive to community credit needs. However, the reasons they gave for why they believed the problems adversely affected their interests—which form the basis for their concerns—and the often contradictory solutions they offered to address the problems, showed that the affected parties differed considerably on how best to revise CRA.

The revised CRA regulations address some, but not all, of the major problems. In response to the first problem, the regulations adopt a results-based examination process. The regulators' success in lessening problems related to inconsistent examinations largely depends on how effectively examiners exercise their discretion when implementing the reforms. To alleviate concerns about insufficient information, the regulations clarify the data to be used to assess results against performance-based standards. However, the affected parties disagree about whether the data collection requirements provide for meaningful performance assessment or are unduly burdensome. The regulations do not address the different enforcement concerns of bankers and community groups.

From its review, GAO believes that some of the difficulties that have hindered past CRA implementation efforts will likely continue to challenge the regulators as they implement the revised regulations. These difficulties include (1) differences in examiner training and experience levels as well as differences in how examiners interpret vague CRA standards; (2) insufficient information to assess institutions' CRA performance and inadequate disclosure in public evaluation reports of the information and rationale used to determine institutions' CRA performance ratings; and (3) insufficient time for examiners to complete all of their responsibilities during CRA examinations. In addition, some regulators were unable to complete CRA examinations for all their banks within their proposed time frames. Furthermore, the regulators estimate that the revised regulations will increase examiner responsibilities, including performing analyses previously required of institutions.

GAO also found from its review that, independent of the regulatory and legislative reform efforts, many bankers, regulators, community groups, and others have taken part in a variety of individual and cooperative initiatives to improve institutions' community lending and reduce related burdens. Through these initiatives, according to participants, institutions have been able to overcome real or perceived barriers to lending in low- and moderate-income areas (community lending). Barriers to community lending and investment may include a variety of economic factors, such as higher costs and risks of community lending compared with other lending and underwriting requirements of major participants in the secondary mortgage markets.

Regulators, to varying degrees, have also played a key role in facilitating cooperation and disseminating information to their institutions about such initiatives through outreach efforts of their community affairs programs. As they further develop these programs and better coordinate their efforts, the regulators' role in this respect should be enhanced.

Congress has considered proposals to amend CRA to reduce the compliance burden and to exempt small institutions from its requirements. In addition, Congress, in recently enacted legislation, has encouraged community development lending. Further, other legislation has been proposed to encourage community lending through financial subsidies or other positive incentives.

INDEX

ABOUT THE EDITORS

June Manning Thomas is Professor of Urban and Regional Planning at Michigan State University, with a joint appointment in Urban Affairs Programs. She is author of *Redevelopment and Race: Planning a Finer City in Postwar Detroit* (1997). Her article, "Planning History and the Black Urban Experience: Linkages and Contemporary Implications," won the Gold Award for the best feature article published in 1994 among scholarly journals associated with the Society of National Association Publications.

Marsha Ritzdorf is Associate Professor of Urban Affairs and Planning at Virginia Polytechnic and State University. She is best known for her historical, empirical, and theoretical work on the race, class, and gender impacts of zoning and land use planning.

ABOUT THE CONTRIBUTORS

Robert A. Catlin is Dean of the Faculty of Arts and Sciences and Professor of Urban Planning at Rutgers University, Camden. A practicing urban planner since 1961 and an academic since 1972, his research interests include community and economic development, housing, planning history/theory and planning issues in minority group communities. He is the author of *Racial Politics and Urban Planning: Gary, Indiana 1980-1989*, published in 1993 by the University of Kentucky Press.

Robert W. Collin is Associate Professor of Environmental Studies at the University of Oregon. He has taught at the University of Virginia, Cambridge University in Cambridge, England, and the University of Auckland. He has published widely in a variety of journals and published one of the first environmental justice law review articles.

Robin Morris Collin is Associate Professor at the University of Oregon Law School. She has also taught at Tulane University Law School, McGeorge School of Law, and Washington and Lee University Law

School. Her areas of interest and expertise include art law, cultural property, environmental law, and sustainability, and she has taught the first seminar on sustainability in an American law school. She is also an active lecturer to civic and professional organizations.

Charles E. Connerly is Associate Professor of Urban and Regional Planning at Florida State University in Tallahassee. From 1991 through 1996, he was coeditor of the *Journal of Planning Education and Research*. His research areas focus on housing and community development and the history of planning and race.

Norman Krumholz is Professor in the Levin College of Urban Affairs at Cleveland State University, Cleveland, Ohio. Prior to his work as professor, he served several cities as a planning practitioner, including ten years as Cleveland Planning Director (1969-1979). He is a past president of the American Planning Association. His most recent book (with Pierre Clavel) is *Reinventing Cities: Equity Planners Tell Their Stories*.

Jacqueline Leavitt teaches in UCLA's School of Public Policy and Social Welfare in the Faculty of the Department of Urban Planning. She is the author of a monograph, *Defining Cultural Differences in Space: Public Housing as a Microcosm*, and coeditor of *The Hidden History of Housing Cooperatives*. Her applied research focuses on public housing and gender.

Raymond A. Mohl is Professor of History and Chair of the Department of History at the University of Alabama at Birmingham. He is the author or editor of ten books on U.S. urban history, including *The New City: Urban America in the Industrial Age, 1860-1920* (1985); *Urban Policy in Twentieth-Century America* (1993); and *The New African American Urban History* (1996). He is currently completing a book on the history of race relations in metropolitan Miami and a study of the second ghetto in mid-twentieth-century America.

Yale Rabin is Professor of Planning Emeritus at the University of Virginia. His fieldwork, research, and publications for over thirty years have concerned the impacts of the land use and housing policies of government on low-income minority groups. He has conducted case studies in over sixty U.S. cities and towns. He has been a consultant to the U.S. Department of Housing and Urban Development, the U.S. Commission on Civil Rights, state and local governments, and numerous national and local legal defense and law reform organizations.

Siddhartha Sen is Assistant Professor in the Graduate Program in City and Regional Planning at Morgan State University in Baltimore. He has published several articles and book chapters on international planning and urban design.

Sigmund C. Shipp is Assistant Professor in the Department of Urban Affairs and Planning at Hunter College in New York City. His research has focused on Black business development. Currently, he is completing a series of papers on Black cooperatives.

Christopher Silver is Professor in the Department of Urban Studies and Planning at Virginia Commonwealth University, Richmond, Virginia. He is author of *Twentieth Century Richmond: Planning, Politics, and Race* (1984); with John V. Moeser, *The Separate City: Black Communities in the Urban South, 1940-1968* (1995); and, with Mary C. Sies, *Planning the Twentieth-Century American City* (1996). Coeditor of the *Journal of the American Planning Association,* he is also serving, on leave from Virginia Commonwealth University, as Urban Development Adviser to the Indonesian government's National Development Planning Agency in Jakarta (1995-1998).

Bobby Wilson is Associate Professor of Geography and Public Affairs at the University of Alabama at Birmingham and a Research Fellow at the Center for Urban Affairs. He is a member of the American Institute of Certified Planners, and his research interests include political economy of race and planning.